The Wit & Wisdom of Winston Churchill

James C. Humes

The Wit & Wisdom of
Winston Churchill

✥ A TREASURY OF MORE THAN
1,000 QUOTATIONS AND ANECDOTES

WITH A FOREWORD BY

Richard M. Nixon

HarperPerennial
A Division of HarperCollins*Publishers*

A hardcover edition of this book was published in 1994 by HarperCollins Publishers.

THE WIT AND WISDOM OF WINSTON CHURCHILL. Copyright © 1994 by James C. Humes. All rights reserved. Printed in the United States of America. No part of this book may be used or reproduced in any manner whatsoever without written permission except in the case of brief quotations embodied in critical articles and reviews. For information address HarperCollins Publishers, Inc., 10 East 53rd Street, New York, NY 10022.

HarperCollins books may be purchased for educational, business, or sales promotional use. For information, please write: Special Markets Department, HarperCollins Publishers, Inc., 10 East 53rd Street, New York, NY 10022.

First HarperPerennial edition published 1995.

Designed by Barbara DuPree Knowles

The Library of Congress has catalogued the hardcover edition as follows:

Churchill, Winston, Sir, 1874-1965.
    The wit and wisdom of Winston Churchill : a treasury of more than 1000 quotations and anecdotes / [compiled by] James C. Humes ; with a foreword by Richard M. Nixon. — 1st ed.
      p. cm.
    Includes bibliographical references (p.) and index.
    ISBN 0-06-017035-2
    1. Churchill, Winston, Sir, 1874–1965 — Quotations.   2. Quotations, English.
    3. Anecdotes.   I. Humes, James C.   II. Title.
DA 566.9.C5A25   1994
941.084′092 — dc20   93-5360

ISBN 0-06-092577-9 (pbk.)
95 96 97 98 99 HC 10 9 8 7 6 5 4 3 2 1

To Martin Gilbert,
whose epic biography etches indelibly the magnitude of Churchill.

To the International Churchill Society and its leader Richard Langworth,
for preserving the memory of Churchill.

Contents

Acknowledgments

It was my mother who first interested me in Winston Churchill. When I was five, she made me listen to some of his war broadcasts. In fact, my first trousers were a replica of a gray-belted Royal Air Force uniform. I wore proudly on my lapel an RAF pin that my mother had earned for her work for Bundles for Britain in 1940.

My mother also insisted that I accept an English-Speaking Union Exchange Scholarship to a British public school in 1952. (I had wanted to go that fall to Williams College, which my brothers had attended.) While at Stowe School, I had the chance to meet Churchill (as well as dance with the Queen), the result of which was a terminal case of Anglophilia.

In my brief meeting with Churchill in May 1953, the prime minister told me, "Young man, study history, study history. In history lie all the secrets of statecraft."

I did study history, and particularly English history and Churchill. The first books on Churchill I read were gifts from my mother in 1953. They were Robert Lewis Taylor's *Churchill: An Informal Biography* and Violet Bonham-Carter's *Winston Churchill*. Both stimulated a college freshman's fascination with Churchill.

Yet it was my years as a White House speechwriter that prompted me to write *Churchill: Speaker of the Century*, published in 1980. Churchill employed no speechwriters, and that fact triggered my desire to read all of his addresses.

For that book and this one I relied on Robert Rhodes James for his eight-volume compilation of all of Churchill's addresses from 1897 to 1963.

The publication of my Churchill biography led the English-Speaking Union to organize a lecture tour on my behalf.

Speeches to sixty-five branches in the United States—as well as others in Canada, the United Kingdom, Australia, and New Zealand—established me as a lecturer on Churchill and brought me in contact with some of his friends and former associates.

To the English-Speaking Union, its former executive director, John Walker, and its current president, Sinclair Armstrong, and their staffs in New York and London, I owe an incalculable debt. Without the English-Speaking Union, I would not have attended a British public school, I would not have met Churchill, and I would not have become a lecturer on Churchill throughout the world. I am also indebted to my friend Kay Halle, who was in turn a friend of the Churchill family. Her *Irrepressible Churchill* is a must for those who seek his wit and humor. I also want to thank Eileen and Howard Lund, my English "parents," whose Holywich House is close to Chartwell, which they used to take me to in my visits with them over the years.

To some it might have been a labor to read all of Churchill's works—his books, his histories, his essays, and his columns. For me it was a delight. I especially recommend the reading of *My Early Life* for glimpses of the Churchill personality.

Other biographies, in addition to Violet Bonham-Carter's, which offer valuable insights into Churchill as a person are Princess Bibescu's *Sir Winston Churchill, Master of Courage* and the diaries of Churchill's doctor, Lord Moran.

Of course, Martin Gilbert's eight-volume work is the monumental biography of the most towering figure of our time. I particularly recommend that Churchill aficionados do not neglect his supplemental companion appendixes for those volumes, which are replete with nuggets of Churchill lore.

I take pride that former President Nixon offered to write the Foreword. I have had the unique experience of working, in various capacities, usually speechwriting, for Dwight Eisenhower, Richard Nixon, Gerald Ford, Ronald Reagan, and George Bush. Of all those leaders, Richard Nixon was the most brilliant and intellectually stimulating. In his world vision and mastery of foreign policy statecraft he comes the closest to the Churchill genius of any American president.

I also thank my editor, Stephanie Gunning, for her patience and suggestions, and my typists, Ahnee Voelker and Peg Whalen. Any errors are due to me and not to them.

Finally, I thank him who shook my hand on that May day in London and told me to "study history."

Foreword

Winston Churchill prophesied, made, and recorded history. In the chronicles of world events, it is difficult to think of others—besides Churchill—who qualify for this singular distinction.

As a young subaltern, he rode in one of the last cavalry charges, in the Sudan in 1895. As a British prime minister, he was consulted when the United States launched the hydrogen bomb in 1953. He served in the cabinet in two world wars. When he first took his seat in Parliament in 1900, he took his oath to Queen Victoria. When he resigned his seat in Parliament in 1964, Lyndon Johnson was U.S. President.

The number of years Churchill served in Parliament and high office outdistances anyone in history. Yet the number of words he created and drafted as author, historian, and journalist, as well as speaker in the House of Commons and for radio addresses and public occasions, also exceed that of almost any writer in this century.

James Humes has selected from this enormous volume close to a thousand excerpts of Churchill's insight and wisdom. With the voice and authority of his experience, Churchill offers invaluable and indispensable lessons.

Churchill is one of the few statesmen who occupied both the world of thought and the world of action.

Most of Churchill's observations carry with them both the dimensions of actual experience and the knowledge of history.

Churchill once wrote of Prime Minister Lord Rosebery that he was "a great man in an era of small events." In an era of momentous events, Churchill was a giant.

Any student of politics, aspiring world leader, or lover of history will find this wit and wisdom of Churchill both a pleasure and an education, and will come to a clearer understanding of why Churchill was the "largest human being of our time."

Richard M. Nixon

Introduction

Just before the coronation of Queen Elizabeth in June 1953, Winston Churchill attended the Commonwealth Banquet in London. The prime minister, then in his eightieth year, was introduced to an eighteen-year-old American student. Churchill told the young man:

"Study history, study history—in history lie all the secrets of statecraft."

Well, I was that young English-Speaking Union exchange scholar in London that day and I did study history—and some of that history was devoted to the study of Churchill. This represents my third Churchill-related book. In addition, I have lectured on Churchill in all fifty states and eighteen countries, including the Commonwealth nations of Canada, New Zealand, Australia, and Malaysia, as well as Bermuda and Hong Kong.

After I would deliver my address, at least one in the audience would take me aside and impart—as if it were a special private secret—that Churchill was his or her favorite hero. Among those who confided their love of Churchill to me have been U.S. senators, Supreme Court justices, prime ministers, Cabinet secretaries and ministers both in the United States and abroad, college presidents,

CEOs of Fortune 500 companies, star athletes, and symphony orchestra conductors.

Why is Churchill the special hero of Fritz Hollings, Jack Kemp, and Ted Williams?

In Philadelphia, a Ugandan waitress told me the only framed picture in the house where she grew up was that of Churchill. As a girl of five, she used to wonder why her father kept the picture of that fat white man. This story did not surprise me, for in a village in Papua New Guinea I had seen another photograph of Churchill—taken from a magazine—in an honored niche.

In 1964, a nine-year-old girl from Colombia, South America, crafted a birthday card. The card, which she addressed only "To the Greatest Man in the World," arrived without stamps at 28 Hyde Park Gate in time for Churchill's ninetieth birthday. He was the giant of our age.

What is Churchill's enduring claim as the century's dominant hero and personality? Certainly his span of influence in the historical arena is unchallenged. When he was elected to Parliament in 1900, he took an oath to Queen Victoria. In 1964, he chose at age eighty-nine not to stand again for his constituency in Woodford, a suburban town outside London, and Queen Elizabeth expressed her regret. He served in the cabinets presiding over two major world wars. As a young member of Parliament he was invited to the White House by President McKinley in 1900 and introduced in a speech by Mark Twain. In 1963 he would accept honorary U.S. citizenship from President Kennedy.

Churchill—like Leonardo da Vinci or Benjamin Franklin—was a Renaissance man. His multitudinous talents shone splendidly. Churchill was a soldier, journalist, author, artist, sportsman, historian, orator, statesman, inventor, and stonemason.

As a soldier, he saw action in four continents—Europe, Asia, Africa, and Latin America. Three times he narrowly escaped death.

As an author, he has had more words published than Dickens and Sir Walter Scott combined and his royalties as a writer exceeded those of Hemingway and Faulkner. In an age when almost all politicians had their major addresses crafted by speechwriters, Churchill wrote his own. In fact, his wartime speeches belong to the realm of literature, and for them he won a Nobel Prize.

As a historian, he authored many volumes recounting the struggles of two world wars and later the epic *History of the English-Speaking Peoples*, as well as biographies of his father, Lord Randolph, and his ancestor the Duke of Marlborough.

As a sportsman, he was a ranking English polo player until age fifty. A fine shot, Churchill hunted rhinoceros in Africa, grouse in Scotland, and quail in Italy. In the 1950s he even stabled a champion racehorse named Colonist II.

As a man of ideas, he conceived of the use of the tank to break the stalemate of trench warfare in World War I. Then, in World War II, he pioneered the idea of the portable landing harbors for the D-Day invasion of France.

As an artist, he employed a heavy palette to fashion impressionist-style paintings. His landscapes won him entry into the Louvre under the *nom de brush* Charles Morin. Picasso, who detested Churchill's conservative politics, said that, had the British politician stuck to painting, he might have entered into the top echelon of artists.

And, as a paid-up trades union member, Churchill, as a stonemason, built walls for Chartwell, his country manor house.

As an orator, Churchill has no peer. General Omar Bradley said, right after the Allied victory in Europe, "His speeches were worth an army." John Kennedy stated, in his words accompanying the granting of U.S. citizenship, "He mobilized the English language and sent it into battle." The historian Arnold Toynbee in 1948 concluded that his wartime speeches spelled the difference between survival and defeat for Britain.

Like an Old Testament patriarch, Churchill also stands supreme as the century's preeminent prophet. His "days of the Locust" warnings—during his wilderness years of the 1930s—about the rise of Nazi Germany come to mind. Furthermore, his "Iron Curtain" address in Fulton, Missouri, in 1946 alerted the West to the threat of the Communist Soviet Union. Not even the "Philippics" of Demosthenes were more prescient or more powerful.

In addition to such massive insights, many of his other predictions—in his writings and speeches—came to pass as well: He forecast the eight-hour workday in 1901, the dominance of air power in the next world war in 1917, the superbomb in 1917, and the energy crisis in 1928.

No less a factor in his hold on public affections was Churchill's quick wit, his impish humor, and the robust exuberance of his personality. More anecdotes are told about Churchill than any other figure in history.

I think Churchill himself explained the key to his unique appeal as a heroic giant. When he returned to his old school Harrow in 1941, he told the boys, "Never, never, never, never, never give in—except to dictates of conscience and duty." It is this defiant resolve—despite all odds—against the foes of freedom that will live forever in men's minds.

Those words of Churchill may epitomize his legacy to the world, but it is a daunting task to encapsulate his manifold greatness in a single book. Even Martin Gilbert, the official Churchill historian, told me he felt cramped in reducing his monumental biography to eight volumes and all the companion appendixes.

As a former presidential speechwriter, I can tell you that no authority—not even Lincoln—is cited more by presidents. In these quotations, the would-be speaker can find reinforcement and sturdy authority for almost every issue or problem.

Churchill's great wartime addresses mark him—like Shakespeare—as a supreme craftsman of the English language. There is much to admire in the sublime phrase and rolling cadence of his oratory.

No other personality in history has spawned so many delicious tales. His love of his wife, Clementine; his affection for his poodle, Rufus, and his cats; and his addictions to whiskey, champagne, and cigars are all part of the Churchill legend. I've tried to include, in the anecdotes and quotations that follow, the best of his quick rejoinders, his pinpricks of the pompous, and his rollicking zest for mischief and adventure.

To know Churchill is to be in touch with the epic personality of our time. Churchill is bigger than life and the range of his abilities and achievements is just as large. For any student of history or lover of language, there is no greater subject than that of Winston Churchill.

The Wit & Wisdom of Winston Churchill

Observations and Opinions

Next to the Bible and Shakespeare, Churchill is the most frequent source of quotations. Like Shakespeare, he was a supreme master of the English language and was prolific in his writings.

Shakespeare died at age fifty-two and the concentrated verse of his thirty-seven plays left an immeasurable legacy. Churchill, however, in his sixty-five years in Parliament, left eight vast tomes just of his speeches—and those do not include his many-volumed histories of two world wars and the English-speaking world. In addition, there are the two biographies, of his ancestor, the Duke of Marlborough, and his father, Lord Randolph Churchill, his early autobiographical adventures, his novel, and books encapsulating many of his columns as a journalist.

Of course, in that massive output not every sentence is a crafted gem, but no public papers of any man in history have ever afforded so many wise epigrams, incisive observations, and pungent wit as those of Churchill.

Like Benjamin Franklin—another historic personality with multitudinous talents—Churchill was blessed with a robust sense of humor. He had an acute sense of the foibles of man—the ambitious, the craven, and the pompous. He could also laugh at himself.

No inhibitions bridled this colossal personality. He freely expressed his tastes and opinions on everything from alcohol to Zionism.

As a young officer at the beginning of his career in public service, he read and reread in his barracks his *Bartlett's Familiar Quotations*. Seven decades later, his words alone could spawn a special edition of that great quotation classic.

ACTION

❧ I never worry about action, but only about inaction.

❧ If you travel the earth, you will find it is largely divided into two classes of people—people who say "I wonder why such and such is not done" and people who say "Now who is going to prevent me from doing that thing?"

ADVENTURE

❧ Foolish perhaps but I play for high stakes and given an audience there is no act too daring or too noble.

ADVERSARY

❧ If you cannot best your strongest opponent in the main theater nor he best you; or if it is very unlikely that you do so and if the cost of failure will be very great, then surely it is time to consider whether the downfall of your strongest foe cannot be accomplished through the ruin of his weakest ally, or one of his weaker allies, and in this connection, a host of political, economic and geographical arguments play their part in the argument.

❖ Who can tell how weak the enemy may be behind his flaming front and brazen mask? At what moment will his willpower break?

❖ The short road to ruin is to emulate . . . the methods of your adversary.

ADVERTISING
❖ If we are to supply the needs of the modern world it can only . . . be done by publicity . . . and advertising.

AGE
❖ Youth is for freedom and reform, maturity for judicious compromise, and old age for stability and repose.

AIR POWER
❖ One cannot doubt that flying . . . must in the future exercise a potent influence, not only in the habits of men, but upon the military destinies of states.

❖ The RAF is the cavalry of modern war.

ALCOHOL
❖ All I can say is that I have taken more out of alcohol than it has taken out of me.

❖ When I was younger I made it a rule never to take a strong drink before lunch. It is now my rule never to do so before breakfast.

❖ No one can ever say that I ever failed to display a meet and proper appreciation of the virtues of alcohol.

ALLIANCES

🔸 How much easier is it to join bad companions than shake them off.

🔸 If we are together nothing is impossible, and if we are divided all will fail.

ALLIES

🔸 There is only one thing worse than fighting with allies and that is fighting without them.

AMBASSADORS

🔸 The zeal and efficiency of a diplomatic representative is measured by the quality and not by the quantity of information he supplies.

AMBITION

🔸 Ambition, not so much for vulgar ends, but for fame, glints in every mind.

AMERICA

🔸 How heavily do the destinies of this generation hang upon the government and people of the United States.[1]

🔸 The United States is like a gigantic boiler. Once the fire is lit under it, there is no limit to the power it can generate.

🔸 The Americans took but little when they emigrated except

---

[1] Written in a newspaper column in 1934.

what they stood up in and what they had in their souls. They came through, they tamed the wilderness, they became "a refuge for the oppressed from every land and clime."

❧ The American eagle sits on his perch, a large strong bird with formidable beak and claws.

❧ There are no people in the world who are so slow to develop hostile feelings against a foreign country as the Americans and there are no people who once estranged, are more difficult to win back.

ANECDOTES
❧ Anecdotes are the gleaming toys of House of Commons history.

ANALOGY
❧ Apt analogies are among the most formidable weapons of the rhetorician.

ANGLO-AMERICAN ALLIANCE
❧ Law, language and literature unite the English-speaking world.

❧ I am myself an English-speaking Union.

❧ You [America] may be the larger and we [Britain] may be the older. You may be the stronger, sometimes we may be the wiser.

❧ Bismarck once said that the supreme fact of the 19th century was that Britain and the United States spoke the same language. Let us make sure that the supreme fact of the 20th century is that they tread the same path.

❧ I read the other day that an English nobleman . . . stated that

England would have to become the 49th state. I read yesterday that an able American editor had written that the United States ought not to be asked to re-enter the British Empire. . . . [T]he path of wisdom lies somewhere between these scarecrow extremes.[2]

❧ It is not given to us to peer into the mysteries of the future. Still I avow my hope and faith, sure and inviolate, that in the days to come, the British and American people will for their own safety and for the good of all walk together in majesty, in justice, and in peace.

ANIMALS
❧ Dogs look up to you, cats look down on you. Give me a pig. He just looks you in the eye and treats you as an equal.[3]

ANTI-APPEASEMENT
❧ If you will not fight for the right when you can easily win without bloodshed, if you will not fight when your victory will be sure, you may come to the moment when you will have to fight with all the odds against you and only a precarious chance of survival.

❧ The sequel to the sacrifice of honor would be the sacrifice of lives, our people's lives.

❧ We seem to be very near the bleak choice between war and shame. My feeling is that we shall choose shame and then have war thrown in a little later on even more adverse terms than at present.

---

[2] Said in a speech in New York in 1946.
[3] Churchill, a lover of animals, included in his menagerie, at one time or another, birds, lambs, and pigs as well as cats and dogs. Two pet swans patrolled his pond and he kept his African lion, Rota, in the zoo. He also raised and rode horses.

APPEASEMENT

✢ An appeaser is one who feeds the crocodile hoping it will eat him last.

✢ It is no use trying to satisfy a tiger by feeding him with cat's meat.

✢ Appeasement in itself may be good or bad according to circumstances. Appeasement from strength is magnanimous and noble and might be the surest and perhaps the only road to world peace.

ARCHITECTURE

✢ We shape our dwellings and afterwards our dwellings shape us.[4]

ARISTOCRACY

✢ Lord Curzon says, "All civilization has been the work of aristocracies." Why, it would be much more true to say the upkeep of civilization has been the hard work of all civilizations.

ARMIES

✢ It is not a case of quantity. . . . Quality rather than quantity; leaders rather than generals; men not masses.

✢ In making an army, three elements are necessary—men, weapons and money. There must also be time.

---

[4] Said in 1944 about the size of the original House of Commons, during the rebuilding after it was bombed.

ART

❧ Art is to beauty what honor is to honesty.

❧ Without tradition art is a flock of sheep without a shepherd. Without innovation it is a corpse.

ARTISTS

❧ Happy are the painters for they shall not be lonely.

ARTS

❧ Ill fares the race which fails to salute the arts with the reverence and delight which are their due.

ASSASSINATION

❧ No nation has ever established its title deeds by assassination.

ATLANTIC CHARTER

❧ The Atlantic Charter was not a law but a star.

AUTHORITY

❧ A supreme enterprise was cast away through my trying to carry out a major and cardinal operation of war from a subordinate position. Men are ill-advised to try such ventures.[5]

AUTOMOBILE

❧ I have always considered that the substitution of the internal combustion engine for the horse marked a very gloomy milestone in the progress of mankind.

---

[5] Churchill was referring to the ill-fated Dardanelles invasion in World War I.

AVOCATION

❧ Change is the master key. A man can wear out a particular part of his mind by continually using it and tiring it, just in the same way as he can wear out the elbows of a coat by rubbing the frayed elbows; but the tired parts of the mind can be rested and strengthened, not by merely rest, but by using other parts. . . . [6] Many men have found great advantage in practicing a handcraft for pleasure. Joinery, chemistry, bookbinding, even bricklaying, if one were interested in them and even skillful at them—would give relief to the tired brain.

BATTLE

❧ Nothing in life is so exhilarating as to be shot at without result.[7]

❧ In all battles two things are usually required of the Commander-in-Chief: to make a good plan for his army and secondly to keep a strong reserve.

❧ War between equals in power should be a succession of climaxes on which everything is staked. These climaxes are called battles. A battle means that the whole of the resources on either side that can be brought to bear during the course of a single episode is concentrated upon the enemy.

❧ I advise you to be in good fettle, as we are in for a good fight.

---

[6] Churchill was explaining how painting refreshed him mentally.
[7] Churchill was referring to being shot at in 1897 with the Makaland Field Force. President Reagan used this quotation as he was carried into the hospital in 1981 after he was shot. The author had supplied him with the words in a toast to Prime Minister Margaret Thatcher the previous January.

BEAUTY

✣ It is hard, if not impossible, to snub a beautiful woman; they remain beautiful and the rebuke recoils.

BIGOTRY

✣ How can any man help how he is born?

BIRTH

✣ It's an extraordinary business this way of bringing babies into the world. I don't know how God thought of it.

BLAME

✣ If the present tries to sit in judgment of the past, it will lose the future.

BLUNTNESS

✣ Do not hesitate to be blunt. . . . This is better done by manner and attitude than by actual words which can be reported.

BOOKS

✣ There is a great deal of difference between the tired man who wants a book to read and the alert man who wants to read a book.

✣ There is a good saying that when a new book appears one should always read an old one.

✣ If you cannot read all your books, at any rate handle, or as it were, fondle them—peer into them, let them fall open where they will, read from the first sentence that arrests the eye, set them back on their shelves with your own hands, arrange them on your own plan so that if you do not know what is in them, you at least know where they are. Let them be your friends, let them at any rate be your acquaintances.

❧ Books in all their variety are often the means by which civilization may be carried triumphantly forward.

BOREDOM
❧ Broadly speaking, human beings may be divided into three classes: those who are billed to death; those who are worried to death; and those who are bored to death.

BRANDY
❧ Good cognac is like a woman. Do not assault it. Coddle and warm it in your hands before you sip it.

❧ I neither want it [brandy] nor need it but I think it pretty hazardous to interfere with the ineradicable habit of a lifetime.[8]

BREAKFAST
❧ My wife and I tried two or three times in the last few years to have breakfast together but it was so disagreeable we had to stop.

❧ Breakfast should be in bed alone.

---

[8] Churchill's choice of cognac was Napoleon, named for the historical figure who most fascinated him.

BRINKMANSHIP

❧ It is no use leading other nations up the garden and then running away when the dog growls.

BRITAIN

❧ We have no assurance that anyone else is going to keep the British Lion as a pet.

❧ It is a curious fact about the British Islanders, who hate drill and have not been invaded for a nearly thousand years, that as danger comes nearer and grows they become progressively less nervous; when it is imminent they are fierce, when it is mortal they are fearless.

❧ I reject the view that Britain and the Commonwealth should now be relegated to a tame and minor role in the world.[9]

❧ Britain is like Laocoön strangled by old school ties.

BUDGET

❧ In finance everything that is agreeable is unsound and everything that is sound is disagreeable.

BULLIES

❧ When nations or individuals get strong they are often truculent and bullying, but when they are weak they become better mannered.

---

[9] This statement by Winston Churchill was read by his son Randolph at the White House in 1963 to acknowledge the honorary U.S. citizenship bestowed upon his father.

BUREAUCRACY

❧ Nothing makes departments so unpopular as acts of petty bureaucratic folly which . . . I fear . . . as only typical of a vast amount of the silly wrongdoing by small officials or committees.

❧ We are indeed victims of a feeble and weary departmentalism.

❧ There is no surer method of economizing and saving money than in the reduction of the number of officials.

❧ A discreditable advertisement of administrative infirmity.

❧ We are not relieving ourselves of burdens and responsibilities. . . . We are merely setting the scene for a complicated controversy, merely creating agencies which will make it more difficult to discharge our task.

CABINETS

❧ Reconstructing a cabinet is like solving a kaleidoscopic jigsaw puzzle.

CANDIDATES

❧ He is asked to stand, he wants to sit, he is expected to lie.

CAMPAIGNING

❧ First of all grin or as they say "smile." There is nothing like it. Next, be natural and quite easy as if you were talking to people in a quiet place about something in which you were much interested. Third, cultivate a marked sense of detachment from the clatter and clamor proceeding around you.

✤ It is no good going to the country solely on the platform of your opponents' mistakes.

CANADA
✤ Canada is the linchpin of the English-speaking world.

CAPITALISM
✤ The vice of capitalism is that it stands for the unequal sharing of blessings; whereas the virtue of socialism is that it stands for the equal sharing of misery.

✤ I am astonished to see how people are afraid to defend the capitalist system. The politicians are afraid, the newspapers are afraid. . . . As a matter of fact the capitalist system is capable of sustained and searching defense.

✤ Is it better to have equality at the price of poverty or well-being at the price of inequality?

CAPITAL PUNISHMENT
✤ I wonder whether in shrinking from the horror of inflicting a death sentence, honorable Members who are conscientiously in favor of abolition do not underrate the agony of a life sentence.

CAUTION
✤ The counsels of prudence and restraint may become the prime agents of mortal danger.

CELEBRITIES

✛ Our aristocracy has largely passed from life into history; but our millionaires, the financiers, the successful pugilists and the film stars who constitute our modern galaxy and enjoy the same kind of privileges as did the outstanding figures of the seventeenth and eighteenth centuries—are all expected to lead model lives.

CHALLENGE

✛ Do not let us speak of darker days, let us speak rather of sterner days.[10]

✛ He who ascends to the mountaintops shall find the loftiest peaks most wrapped with clouds and snow.

✛ Difficulties mastered are opportunities won.

CHAMPAGNE

✛ A single glass of champagne imparts a feeling of exhilaration. The nerves are braced, the imagination is agreeably stirred, the wits become more nimble. A bottle produces a contrary effect.[11]

CHANGE

✛ To improve is to change; to be perfect is to change often.

---

[10] This was his message to his old school Harrow in 1941. He had the words in the school anthem changed from "darker" days to "sterner" days.
[11] Churchill's choice of champagne was Pol Roger. Madame Pol-Roger in France had kept two cases reserved for Churchill hidden from the Nazis.

✜ My views are a harmonious process which keeps them in relation to the current movement of events.

✜ One cannot leap a chasm in two jumps.

CHEESE
✜ Stilton and port are like man and wife. They should never be separated. Whom God has joined together, let no man put asunder.

CHINA
✜ Chiang Kai-shek deserves the protection of your shield but not the use of your sword.[12]

✜ The tail of China is large and will not be wagged.

CHIVALRY
✜ Let it not be thought that the age of chivalry belongs to the past.

CHOICES
✜ The more man's choice is free, the more likely it is to be wise and fruitful not only to the chosen but to the community in which he dwells.

CHRISTIANITY
✜ The flame of Christian ethics is still our best guide . . . only on this basis can we reconcile the rights of the individual with the demands of society.

---

[12] From a 1955 letter in which Churchill warns President Eisenhower not to get involved in an invasion of China.

CHRISTMAS

✣ Christmas is a season not only of rejoicing but of reflection.

✣ We have tonight the peace of the spirit in each cottage home and in every generous heart.

✣ Let the children have their night of fun and laughter. Let the gifts of Father Christmas delight their play.

CIGARS

✣ Of two cigars pick the longest and the strongest.[13]

✣ Smoking cigars is like a falling in love; first you are attracted to its shape; you stay with it for its flavor; and you must always remember never, never, let the flame go out.[14]

CIVILIAN AUTHORITY

✣ It is always dangerous for soldiers, sailors or armies to play at politics.

CIVILIZATION

✣ A state of society where moral force begins to escape from the tyranny of physical force.

✣ When civilization degenerates: our morals will be gone but our maxims will remain.

---

[13] Told to the author by Churchill's son Randolph in 1963.
[14] Churchill's favorite brand of Havanas was Romeo and Juliet. The special "Churchill" size was six and a half inches.

CIVIL RIGHTS

✧ The foundation of all democracy is that the people have the right to vote.

✧ Nothing can be more abhorrent in a democracy than to detain a person or keep him in prison because he is unpopular.

CIVIL SERVANTS

✧ No longer servants, no longer civil.

CLASS WARFARE

✧ We are opposed to class government in every form, whether it be a government of aristocracy or of plutocracy or of the military classes, or of the priest class or of the trade unions.

✧ There is no case for a quarrel between wealth and poverty, it is a quarrel between methods of government and themes of government.

COLD WAR

✧ What we are faced with is not a violent jerk but a prolonged pull.

✧ It is better to have a world united than a world divided; but it is also better to have a world divided than a world destroyed.

✧ Let there be sunshine on both sides of the Iron Curtain; and if ever the sunshine should be equal on both sides the Curtain will be no more.

❧ I am sure we do not want any fingers on the trigger. Least of all, do we want a fumbling finger.[15]

❧ We are firm as a rock against aggression but the door is always open to friendships.

COMMAND
❧ Don't be careless about yourselves—on the other hand not too careful. Live well but do not flaunt it. Laugh a little and teach your men to laugh—get good humor under fire—war is a game that's played with a smile. If you can't smile, grin. If you can't grin, keep out of the way till you can.

COMMITMENT
❧ No one is compelled to serve great causes unless he feels fit for it, but nothing is more certain than you cannot take the lead in great causes as a half-timer.

COMMUNISM
❧ I will not pretend if I had to choose between Communism and Nazi-ism, I would choose Communism. I hope not to be called on to survive in either.

❧ Trying to maintain good relations with a Communist is like wooing a crocodile. You do not know whether to tickle it under the

[15] Churchill was reacting to the Labour party's campaign charge in 1951—"Whose finger do you want on the trigger?"

chin or beat it over the head. When it opens its mouth, you cannot tell whether it is trying to smile or preparing to eat you up.

✤ The Bolshevik is not an idealist who is content to promote his cause by argument or example.

✤ Bolshevism is not a policy; it is a disease.

✤ The day will come when it will be recognized without doubt throughout the civilized world that the strangling of Bolshevism at birth would have been an untold blessing to the human race.

✤ In Russia, a man is called a reactionary if he objects to having his property stolen and his wife and children murdered.

✤ It's no use arguing with a Communist. It is no use trying to convert a Communist or persuade him . . . you can only do it by having a superior force on your side on the matter in question.

✤ Communism is a religion . . . Jesuits without Jesus.

✤ A ghoul descending from a pile of skulls.

COMPANIONSHIP
✤ This is the company I should like to find in heaven. Stained perhaps, but positive. Not those flaccid sea anemones of virtue who can hardly wiggle an antenna in the turgid waters of negativity.

COMPETITION
✤ You have taken the measure of your foe; you have only to go forward with confidence.

CONFERENCES

❧ Policy first. Atmosphere second. Then and not till then—Action.[16]

CONFLICT

❧ Great quarrels . . . arise from small occasions but seldom from small causes.

❧ Arm yourselves and be ye men of valor and be in readiness for the conflict; for it is better for us to perish in battle than to look upon the outrage of our nation and our altar.

❧ The worst quarrels only arise when both sides are equally in the right and in the wrong.

CONSCIENCE

❧ The final tribute is our own conscience.

❧ The only guide to a man is his conscience; the only shield to his memory is the rectitude and sincerity of his actions.[17]

CONSERVATISM

❧ The difference between our outlook and the Socialist outlook is the difference between the ladder and the queue. We are for the ladder. Let all try their best to climb. They are for the queue.

---

[16] This comment is about the Versailles Peace Conference in 1919.
[17] Eulogy of Neville Chamberlain in 1940.

✤ The Conservative Party stands for a way of life which at every stage multiplies the choice open to the Socialist devotees. . . . We plan for choices, they plan for rules.

✤ Our conservative principles are well known. We stand for the free and flexible working of the supply and demand. We stand for compassion and aid for those who, whether through age, illness or misfortune, cannot keep pace with the march of society.

✤ The Socialists aim at the maximum of regulations and the Conservatives aim at the minimum.

✤ Do not let spacious plans for a new world divert your energies from saving what is left of the old.

✤ The only path to safety is to liberate the energies and genius of the nation and let them have their full fruition.

✤ We must beware of trying to build a society in which no one counts for anything except a politician or an official, or a society where enterprise gains no reward and thrift no privileges.

✤ We shall return to a system which provides incentives for effort, enterprise, self-denial, initiative and good housekeeping. We cannot uphold the principle that the rewards of society must be equal for those who try and those who shirk, for those who succeed and for those who fail.

CONSISTENCY
✤ The only way a man can remain consistent amid changing circumstances is to change with them while preserving the same dominating purpose.

CONSULTATION

❧ Leaders who lead their party from day to day by doing the popular thing, by staving off difficulties and by withholding their time course until it is too late, cannot complain if, when the disaster culminating in catastrophe is reached, some of their followers are reluctant to share in the odium of capitulation.

❧ Well, one can always consult a man and ask him, "Would you like your head cut off tomorrow?" and after he has said "I would rather not," cut it off. "Consultation" is a vague and elastic term.

COOPERATION

❧ There are two ways of securing cooperation in human action. You get cooperation by controls or you can get it by comprehension.

COURAGE

❧ Courage is rightly esteemed the first of human qualities . . . because it is the quality which guarantees all others.

❧ No one in great authority had the wit, ascendancy or detachment from public folly to declare these fundamental, brutal facts to the electorate.[18]

CRISIS

❧ The fact that a number of crises break out at the same time does not necessarily add to the difficulty of coping with them. One set of adverse circumstances may counter-balance and even cancel out another.

---

[18] Churchill was denouncing the appeasement policies in the 1930s.

CRITICISM

✥ Although always prepared for martyrdom, I preferred that it should be postponed.

✥ Criticism is easy; achievement is difficult.

✥ Eating my words has never given me indigestion.

✥ I am always ready to learn, although I do not always like being taught.

✥ Criticism in the body politic is like pain in the human body. It is not pleasant but where would the body be without it?

CRITICS

✥ It is not open to the cool bystander . . . to set himself up as an impartial judge of events which would never have occurred had he outstretched a helping hand in time.

DANGER

✥ Our difficulties and our dangers will not be removed by closing our eyes to them.

✥ Thought arising from a factual experience may be a bridle or a spur.

DEATH

✥ The world does not end with the life of any man.

✥ When the notes of life ring false, men should correct them by referring to the tuning fork of death.

⚜ Any man who says he is not afraid of death is a liar.

⚜ Black velvet . . . eternal sleep.

⚜ We make too much of it. All religions do. Of course, I may alter my views.

⚜ Old age, by blanching the seat of reason, may cut off the fear of death even in a once imaginative mind, or it may, on the other hand, undermine fortitude, softening the will.

DECISION MAKING
⚜ More difficulty and toil are often incurred in overcoming opposition and adjusting divergent and conflicting views than by having the right to give decisions itself.

DECLARATION OF INDEPENDENCE
⚜ The Declaration of Independence is not only an American document. It follows on the Magna Charta and the Bill of Rights as the third great title-deed in which the liberties of the English-speaking people are founded.

DECLINE
⚜ I have not become the King's First Minister in order to decline over the liquidation of the British Empire.[19]

---

[19] Said in 1943.

DEFEAT

✢ Although always prepared for martyrdom, I prefer that it shall be postponed.

✢ There is only one answer to defeat and that is victory.

✢ Defeat is one thing; disgrace is another.[20]

DEFENSE

✢ When nations are strong, they are not always just and when they wish to be just, they are no longer strong.

✢ I hope that I shall never see the day when the Force of Right is deprived of the Right of Force.

✢ The Romans had a maxim. "Sharpen your weapons and limit your frontiers." But our maxim seems to be "Diminish your weapons and increase your obligations."

✢ I cannot subscribe to the idea that it might be possible to dig ourselves in and make no preparations for anything else than passive defense. It is the theory of the turtle.

DEFICIT

✢ There are two ways in which a gigantic debt may be spread over new decades and future generations. There is the right and healthy way; and there is the wrong and morbid way. The wrong way is to fail to make the utmost provision for amortisation which prudence al-

---

[20] Said about the British defeat at Tobruk in Egypt in 1942.

lows, to aggravate the burden of the debt by fresh borrowing, to live from hand to mouth and from year to year, and to exclaim with Louis XVI: "After me, the deluge!"

DELAY

✣ A policy which hopes to succeed by postponing occasions can . . . hardly hope to resist the whirlwind.

✣ If we go on waiting upon events, how much shall we throw away our resources now available for our security?

DELIBERATION

✣ There is, therefore, wisdom in reserving one's decisions as long as possible and until all the facts and forces that will be potent at the moment are revealed.

DEMAGOGUERY

✣ Social distinction was always the highest ambition of the demagogue.

DEMOCRACY

✣ Democracy is no harlot to be picked up on the street by a man with a tommy gun.

✣ Where there is a great deal of free speech, there is always a certain amount of foolish speech.

✣ We depend on the private soldier of the British democracy. We place our trust in the loyal heart of Britain. Our faith is founded upon the rock of the wage-earning population of this island which has never yet been appealed to, by duty and chivalry, in vain.

✤ Democracy is the worst form of government except for all those other forms that have been tried from time to time.

✤ It is the people who control the Government—not the Government the people.

✤ I am a child of the House of Commons. I was brought up in my father's house to believe in democracy.

✤ The genius . . . springs from every class and from every part of the land. You cannot tell where you will not find a wonder. The hero, the fighter, the poet, the master of science, the organizer, the engineer, the administrator or the jurist—he may spring into fame. Equal opportunity for free institution and equal laws.

DEMONSTRATIONS

✤ We may not be heard tonight, but we will carry out our purpose at the poll. We will not submit to the bullying tyranny of the featherheads. We will not submit to the roar of the mob.

✤ We have a most valuable right, that democracy shall be able to conduct its public gatherings with decency and order.

✤ Chivalrous gallantry is not among the peculiar characteristics of excited democracy.

DEPRESSION

✤ Do not let us aggravate industrial depression by the undermining of credit.

DESPAIR

✢ It is a crime to despair. We must learn to draw from misfortune the means of future strength.

✢ Never flinch, never weary, never despair.

✢ There is one cardinal rule: "Never Despair." That word is forbidden.

DESTINY

✢ Our future is in our hands. Our lives are what we choose to make them.

✢ My conclusion on Free Will and Predestination . . . they are identical.

DICTATORS

✢ Dictators ride to and fro upon tigers from which they dare not dismount. And the tigers are getting hungry.

✢ When heads of states become gangsters, something has to be done.

✢ One of the disadvantages of dictatorships is that the dictator is often dictated to by others, and what he did to others may often be done back again to him.

✢ You see these dictators on their pedestals surrounded by the bayonets of their soldiers and truncheons of their police. . . . [T]hey boast and vaunt themselves before the world, yet in their hearts, there is unspoken fear. They are afraid of words and thoughts stirring at home—all the more powerful because forbidden—terrify them. A little mouse, a tiny little mouse of thought appears in the room and even the mightiest potentates are thrown into panic.

DICTATORSHIP

✤ The monstrous child of emergency.

✤ The fetish worship of one man.

DINNER

✤ My idea of a good dinner is first to have good food, then discuss good food, and after this good food has been elaborately discussed, to discuss a good topic—with myself as the chief conversationalist.

DIPLOMACY

✤ The reason for having diplomatic relations is not to confer a compliment but to secure a convenience.

✤ Diplomacy is the art of telling plain truths without giving offense.

✤ When you have to kill a man, it costs nothing to be polite.

DISARMAMENT

✤ The cause of disarmament will not be attained by mush, gush and slush—it will be advanced by the harassing expense of fleets and armies.

✤ False ideas have been spread about the country that disarmament means peace.

✣ To remove the causes of war we must go deeper than armaments. We must remove grievances and injustices. . . . Let moral disarmament leave and physical disarmament will follow.

DUTY
✣ Forward then. Forward! Let us go forward without fear into the future and let us dread naught when duty calls.

✣ While we have strength we must discharge our duty. Neither taunts nor blandishments should move us from it.

✣ What we require to do now is to stand erect and look the world in the face and do our duty without fear or favor.

ECONOMIC GROWTH
✣ The process of the creation of new wealth is beneficial to the whole community.

✣ Prosperity, that errant of our daughter who went astray . . . is on our threshold. She has raised her hand to the knocker on the door. What shall we do? Shall we let her in or shall we drive her away?

ECONOMICS
✣ The root problem of modern economics: the strange discordance between the consuming and producing power.

EDUCATION
✣ Those who think that we can become richer or more stable as a country by stinting education and crippling the instruction of our young people are a most benighted class of human beings.

❧ The most important thing about education is appetite.

EGOTISM
❧ Of course, I am an egotist. Where do you get if you aren't?

ENDING
❧ This is not the end. It is not even the beginning of the end. It is perhaps the end of the beginning.[21]

ENFORCEMENT
❧ No more force should be used than is necessary to insure compliance with the law.

ENGLAND
❧ The nose of the bulldog has been slanted backward so that he can breathe without letting go.

ENGLISH LANGUAGE
❧ The essential structure of the ordinary British sentence . . . is a noble thing.

❧ Naturally I am biased in favor of boys learning English and then I would let the clever men learn Latin as an honor and Greek as a treat. But the only thing I would whip them for is not knowing English. I would whip them hard for that.

❧ Let us not shrink from using the short expressive phrase even if it is conversational.

---

[21] Churchill's statement at the victory of El Alamein in Africa in 1942.

※ A vocabulary of truth and simplicity which will be of service through life.

※ Short words are best and old words when short are best of all.

※ English literature is a glorious inheritance in the English language and in its great writers there are great riches and treasures, of which of course the Bible and Shakespeare stand alone on the highest platform.

ENGLISH LITERATURE
※ The English language is one of our great sources of inspiration and strength, and no country, or combination, or power so fertile and so vivid exists anywhere else in the world.

ENGLISH PUBLIC [PRIVATE] SCHOOLS
※ [The purpose of the public schools] is feeding sham pearls to real swine.

ENJOYMENT
※ If this be a world of vice and woe, I'll take the vice and you take the woe.

EUROPE
※ I see no reason why . . . there should not ultimately arise the United States of Europe.[22]

---

[22] Churchill in 1946 voiced his prediction of a European community in an address at the Hague.

✤ There can be no revival of Europe without a spiritually great France and a spiritually great Europe.

✤ We must build a kind of United States of Europe. . . . In this urgent work France and Germany must work together.[23]

EVIL
✤ Wickedness is not going to reign.[24]

EXERCISE
✤ I get my exercise being a pallbearer for those of my friends who believe in regular running and calisthenics.

EXPEDIENCY
✤ Cheap popularity can prove itself very dearly bought.

✤ Nothing is more dangerous . . . than to live in the temperamental atmosphere of a Gallup Poll, always feeling one's pulse and taking one's temperature.

✤ It always looks so easy to solve problems by taking the path of least resistance. What looks like the easy road turns out to be the hardest and most cruel.

✤ There are two kinds of success: initial and ultimate.

---

[23] In the aftermath of the war, Churchill issued this call for a United States of Europe—with France and the just defeated Germany leading the way. Churchill was speaking while accepting an award in Zurich in September 1946.
[24] Churchill's comment about Hitler's invasion of Europe.

✥ To act by half-measures, with a lack of conviction miscalled "caution," is to run the greatest risks and lose the prize.

EXPENDITURE
✥ Expenditure always is popular; the only unpopular part about it is the raising of the money to finance the expenditure.

EXPERTISE
✥ Expert knowledge, however indispensable, is no substitute for a generous and comprehending outlook upon a human story with all its sadness—with all its unquenchable hope.

EXTREMISM
✥ All great movements, every vigorous impulse that a community may feel, may become perverted and distorted as time passes. . . . A wide humanitarian sympathy in a nation easily degenerates into hysteria. A military spirit tends towards brutality, Liberty leans to license, restraint to tyranny. The pride of race is distended to blustering arrogance. The fear of God produces bigotry and superstition.

FACTS
✥ You can't ask us to take sides against arithmetic. You cannot ask us to take sides against the obvious facts of the situation.

✥ Facts are better than dreams.

✥ Never must we despair, never must we give in, but we must face facts and draw true conclusions from them.

✥ You must look at the facts because they look at you.

✢ The butterfly is the Fact—gleaming, fluttering, settling for an instant with wings fully spread to the sun, then vanishing in the shades of the forest.

✢ Only the facts can tell the tale; and the public ought now to have them.

FAITH

✢ The idea that nothing is true except what we comprehend is silly.

✢ It is bad for a nation when it is without faith.

FAMILY

✢ There is no doubt that it is around the family and the home that all the greatest virtues, the most dominating virtues of human society, are created, strengthened and maintained.[25]

✢ You must have four children. One for Mother, one for Father, one for Accidents, one for Increase!

✢ Where does the family start? It starts with a young man falling in love with a girl. No superior alternative has yet been found.

✢ The whole theme of motherhood and family life, with those sweet affections which illuminate it, must be the fountain spring of present happiness and future survival.

---

[25] At the birth of Prince Charles in 1948.

FANATIC

✢ A fanatic is one who won't change his mind and won't change the subject.

FASCISM

✢ Socialism is bad, jingoism is worse and that the two combined in . . . fascism—the worst creed ever designed by man.

FEAR

✢ You will never get to the end of the journey if you stop to shy [toss] a stone at every dog that barks.

FINALE

✢ Let us finish this job in style. We can do it if we want and it is well worth doing.

FLATTERY

✢ Gush, however quenching, is always insipid.

FLEXIBILITY

✢ The best method of acquiring flexibility is to have three or four plans for all the probable contingencies all worked out with the utmost detail.

FLU

✢ It is a nuizenza to have the fluenza.[26]

------

[26] From a letter to President Franklin Delano Roosevelt in 1942.

FLYING

✢ The air is an extremely dangerous mistress. Once under the spell most lovers are faithful to the end, which is not always old age.

FOOD

✢ When one wakes up after daylight one should breakfast; five hours after that, luncheon. Six hours after luncheon, dinner. Thus one becomes independent of the sun, which otherwise meddles too much in one's affairs and upsets the routine of work.

FOREIGN POLICY

✢ In world affairs, it is no use indulging in hate and revenge. They are the most expensive and futile and self-destroying luxuries by which one can squander the treasure accumulated by the valour of your sons and daughters.

✢ Advantage is gained in war and also in foreign policy and other things by selecting from many attractive or unpleasant alternatives the dominating point. . . . Evidently this should be the rule, and other great business be set in subordinate relationship to it. Failure to adhere to this simple principle produces confusion and futility of action, and nearly always makes things much worse later on.

✢ A nation without a conscience is a nation without a soul. A nation without a soul is a nation that cannot live.

FORESIGHT

✢ Dangers which are warded off by effective precaution and foresight are never even remembered.

FRANCE

✢ France though armed to the teeth is pacifist to the core.

FREEDOM

✢ Never surrender ourselves to servitude and shame whatever the cost may be.

✢ The soul of freedom is deathless; it cannot, and will not, perish.

✢ In harsh and melancholy epochs free men may always take comfort from the ground lesson of history that tyrannies cannot last except among servile races.

✢ Laws, just or unjust, may govern men's actions. Tyrannies may restrain or regulate their words. The machinery for propaganda may pack their minds with falsehood. . . . But the soul of man thus held in trance or frozen in a long night can be awakened by a spark coming from God knows where.

FREE MARKETS

✢ If you destroy a free market, you create a black market.

FREE SPEECH

✢ Some people's idea of free speech is that they are free to say what they like but if anyone says anything back, that is an outrage.

FUTURE

✢ It seems to me the tide of destiny is moving steadily in our favor, though the voyage will be long and rough.

✢ I have no fear of the future. Let us go forward into its mysteries, let us tear aside the veils which hide it from our eyes and let us move onward with confidence and courage.

✢ In the past we have a light which flickered; in the present we have a light which flames; and in the future there will be a light which shines over all the land and sea.

✢ It is not given to us to peer into the mysteries of the future.

✢ It is a mistake to look too far ahead. Only one link in the chain of destiny can be handled at a time.

GENERALS
✢ Men have bad luck and their luck may change. But anyhow you will not get generals to run risks unless they feel they have behind them a strong government.

✢ Battles are won by slaughter and maneuver. The greater the general, the more he contributes in maneuver, the less he demands in slaughter.

✢ I rate the capacity of a man to give a useful opinion . . . with the following three conditions: First, courage and ability. Second, real experience of the fire. Third, peace time staff studies and peace time promotions.

✢ I like commanders on land and sea and in air to feel that between them and all forms of public criticism the Government stands like a strong bulkhead. They ought to have a fair chance, and more than one chance.

✢ Generals are often prone, if they have the chance, to choose a set-piece battle, when all is ready, at their own selected moment, rather than to wear down the enemy by continued unspectacular fighting. They naturally prefer certainty to hazard. They forget that war never stops, but burns on from day to day with ever-changing results not only in one theatre but in all.

GENEROSITY
✢ Generosity is always wise.

GERMANS
✢ The Hun is always at your throat or at your feet.

✢ An immense responsibility rests upon the German people for this subservience to the barbaric idea of autocracy. This is the gravamen against them in history—that in spite of all their brains and courage—they worship power and let themselves be led by the nose.

GOD
✢ The glory of man is the glory of God. . . . When I speak of the glory of man, I do not mean the splendor, pomp and pageantry . . . I mean the glory of the man whose soul is at peace within him . . . because he has the peace of well-being in his soul.

GOLF
✢ Golf is like chasing a quinine pill around a pasture.

GRIEF

✤ An old and failing life going out on the tide after the allotted span has been spent and after most joys have faded is not a cause for human pity. It is only part of the immense tragedy of our existence here below from which both faith and hope rebelled.[27]

✤ The span of mortals is short, the end universal; and the tinge of melancholy which accompanies decline and retirement is in itself an anodyne. It is foolish to waste lamentation upon the closing phase of human life. Noble spirits yield themselves willingly to the successively falling shades which carry them to a better world or to oblivion.[28]

HATRED

✤ Hate is a bad guide. I have never considered myself at all a good hater—though I recognize that from moment to moment [hate] has added stimulus to pugnacity.

✤ Hatred plays the same part in government as acid in chemistry.

✤ We must learn the lessons of the past. We must not remember today the hatreds of yesterday.

HISTORY

✤ It is not enough to collect and establish facts, it is not enough to have them checked, corrected and commented upon by experts. What is . . . of supreme importance is how you present them.

---

[27] At the death of his mother.
[28] About the death of Marlborough.

42

⚜ Past experience carries with its advantages, the drawback that things never happen the same way again.

⚜ The Muse of History must not be fastidious.

⚜ We cannot say "the past is past" without surrendering the future.

⚜ Study history, study history—in history lie all the secrets of statecraft.[29]

⚜ A nation that forgets its past has no future.

⚜ History unfolds itself by strange and unpredictable paths. We have little control over the future and none at all over the past.

⚜ A good knowledge of history is a quiver full of arrows in debates.

⚜ Everyone can recognize history when it happens. Everyone can recognize history *after* it has happened; but it is only the wise man who knows at the moment what is vital and permanent, what is lasting and memorable.

⚜ History with its flickering lamp stumbles along the trail of the past trying to reconstruct its success to revive the echoes and kindle with pale gleams the passions of former days.

---

[29] Said to the author when he was an English-Speaking Union exchange scholar in 1953.

❧ Learn all you can about the history of the past, for how else can one even make a guess what is going to happen in the future?

❧ The farther backward you can look, the farther forward you can see.

❧ Those who seek to plan the future should not forget the inheritance they have received from the past, for it is only by studying the past as well as drawing for the future that the story of man's struggle can be understood.

HOLOCAUST

❧ There is no doubt that this is probably the greatest and most horrible crime committed in the history of the world and it has been done by scientific machinery by nominally civilized men in the name of a great state.

HONOREE

❧ I am proud, but I am also awestruck at your decision to include me. I do hope you are right. I feel we are both running a considerable risk and that I do not deserve it. But I shall have no misgivings if you have none.[30]

HOPE

❧ Nourish your hopes but do not overlook realities.

❧ The message of dawn is hope.

---

[30] Churchill's remarks at winning the Nobel Prize in literature in 1953.

- ⚜ There is time and hope if we combine patience with courage.

- ⚜ When you leave off dreaming, the universe ceases to exist.

HORSES
- ⚜ Don't give your child money. As far as you can afford it, give him horses. No hour of life is lost that is spent in the saddle.

HOUSE OF COMMONS
- ⚜ This House is not only a machine for legislation; perhaps it is not even mainly a machine for legislation, it is a great forum of debate. . . . If the House is not able to discuss matters which the country is discussing, which fill the newspapers, which everyone is anxious and preoccupied about, it loses its contact; it is no longer marching step by step with all the thought that is in progress in the country.

- ⚜ What is the use of Parliament if it is not the place where true statements can be brought before the people?

HUMANITY
- ⚜ And I avow my faith that we are marching toward better days. Humanity will not be cast down. We are going on—swinging bravely forward—along the grand high road—and already behind the distant mountains is the promise of the sun.

HUMOR
- ⚜ In my behalf you cannot deal with the most serious things in the world unless you also understand the most amusing.

IDEALISM

❖ The human race cannot make progress without idealism, but idealism at other people's expense . . . cannot be regarded as its highest or noblest form.

❖ Moral force is unhappily no substitute for armed force, but it is a very great reinforcement.

❖ How far, alas, do man's endeavors fall short in practice of his aspirations![31]

IDEALS

❖ We can always try our luck to steer us straight as possible by those harbor lights which cast across the stormy waste of waters— beacons of justice and truth.

❖ In a broad view, large principles, a good heart, high aims and a firm faith we may find some charts and compass for our voyage.

IDLENESS

❖ Idleness is a dangerous breeding ground.

ILLUSIONS

❖ It is no use dealing with illusions and make-believes. We must look at the facts. The world . . . is too dangerous for anyone to be able to afford to nurse illusions. We must look at realities.

---

[31] Churchill's summary of the League of Nations in 1938.

IMAGINATION

✢ The stronger your imagination the more variegated your universe.

✢ Imagination, without deep and full knowledge, is a snare.

✢ No idea is so outlandish that it should not be considered with a searching but at the same time a steady eye.

✢ Energy of mind does not depend on energy of body.

IMPOTENCE

✢ Decided only to be undecided, resolved to be irresolute, adamant for drift, solid for fluidity, all powerful to be impotent.[32]

✢ The finest combination in the world is power and mercy. The worst combination is weakness and strife.

INDIA

✢ India is an abstraction. . . . India is no more a political personality than Europe. India is a geographical term. It is no more a united nation than the Equator.

INDIVIDUALISM

✢ It is not possible to draw a hard-and-fast line between individualism and collectivism. You cannot draw it either in theory or in practice. No man can be a collectivist alone or an individualist

---

[32] Churchill's judgment of the appeasement policies in the 1930s.

alone. He must be both an individualist and a collectivist. . . .
Collectively we light our streets and supply ourselves with water.
But we do not make love collectively and the ladies do not marry us
collectively.

✝ Individualism offers an infinitely graduated and infinitely
varied system of records for genius, for enterprise, for exertion,
for industry, for faithfulness, for thrift. Socialism destroys all
this.

INJURY

✝ The more serious physical wounds are often surprisingly endur-
able at the moment they are received. There is an interval of uncertain
length before sensation is renewed. The shock numbs but does not
paralyze; the wound bleeds but does not smart. So it is with the great
reverses of life.

INNOVATION

✝ We must beware of needless innovations, especially when
guided by logic.

✝ Innovation of course involves experiment. Experiments may or
may not be fruitful.

INTERNATIONAL LAW

✝ Humanity, not legality, must be our guide.[33]

---

[33] Churchill was referring to the Nuremberg trials.

INVITATION

✣ It is a very fine thing to refuse an invitation, but it is a good thing to wait until you get it first.

ISLAM

✣ What the horn is to the rhinoceros, what the sting is to the wasp, the Mohammadan faith is to the Arabs.

ISRAEL

✣ I believe in the idea of creating an autonomous Jewish colony . . . under the flag of toleration and freedom.[34]

✣ A great event is taking place here, a great event in the world's destiny. It is taking place without injury or injustice to anyone; it is transforming waste places into fertile; it is planting trees and developing agriculture in desert lands; it is making for an increase in wealth and of cultivation; it is making two blades of grass from where one grew before.[35]

JARGON

✣ Official jargon can be used to destroy any kind of human contact or even thought itself.

✣ Let us have an end of such phrases as these: "It is also of importance to bear in mind the following considerations . . ." or

---

[34] Churchill voiced these Zionist sympathies in 1906.
[35] As Secretary of State for the Colonies in 1921, Churchill made this statement at a Middle East conference.

"Consideration should be given to the possibility of carrying into effect. . . ." Most of these wooly phrases are mere padding, which can be left out altogether or replaced by a single word.

JOURNALISM
⚜ It is better making the news than [the] taking [down of] it.

JUDGMENT
⚜ It is a fine thing to be honest, but it is also very important to be right.

⚜ What most people call bad judgment is judgment which is different from theirs.

JUDICIARY
⚜ The independence of the judiciary from the executive is the prime defense against tyranny.

JURY
⚜ Trial by jury, the right of every man to be judged by his equals, is among the most precious gifts that England has bequeathed to America.

JUSTICE
⚜ One ought to be just before one is generous.

KNOWLEDGE
⚜ The more knowledge we possess of the opposite point of view, the less puzzling it is to know what to do.

LEADERSHIP

✥ If people were told of their dangers, they would consent to make the necessary sacrifice.[36]

✥ One mark of a great man is the power of making lasting impressions upon the people he meets.

✥ The first trials of a Prime Minister are often the most severe. The most formidable obstacles lie at the beginning. Once these have been surmounted, the path is comparatively smooth.

✥ There is a precipice on either side of you—a precipice of caution and a precipice of over-daring.

✥ When the eagles are silent, the parrots begin to jabber.

✥ People who are not prepared to do unpopular things and defy clamor of the multitude are not fit to be ministers in times of difficulty.

LEFTISTS

✥ They are the most disagreeable of people. . . . Their insincerity? Can you not feel a sense of disgust at the arrogant presumption of superiority of these people? Superiority of intellect! Then, when it comes to practice, down they fall with a wallop not only to the level of ordinary human beings but to a level which is even far below the average.

---

[36] This was said about rearming in the late 1930s in face of the Hitler threat.

❧ Thoughtless, dilettante or purblind worldlings sometimes ask us "What is it that Britain and France are fighting for?" To this I answer "If we left off fighting, you would soon find out."

❧ The worst difficulties from which we suffer do not come from without. They come from within. . . . They come from a peculiar type of brainy people, always found in our country, who if they add something to our culture, take much from its strength. Our difficulties come from the mood of unwarrantable self-abasement into which we have been cast by a powerful section of our own intellectuals.

❧ These very high intellectual persons who wake up every morning . . . see what they can find to demolish, to undermine or cast away.

❧ Collective ideologists—those professional intellectuals who revel in decimals and polysyllables.

❧ We are still a great nation. It is not too late to stem these subversive and degenerating tides you see working and flowing in every direction.

❧ Let them quit these gospels of envy, hate and malice. Let them eliminate them from their politics and programmes. Let them abandon the utter fallacy, the grotesque, erroneous fatal blunder of believing that by limiting the enterprise of man, by riveting the shackles of a false equality . . . they will increase the well-being of the world.

LEGISLATURE

⚜ The practice of Parliament must be judged by quality, not quantity. You cannot judge the passing of laws by Parliament as you would judge the output of an efficient Chicago bacon factory.

⚜ The congestion of Parliament is a disease, but the futility of Parliament is a mortal disease.

⚜ The object of Parliament is to substitute argument for fisti-cuffs.

⚜ It is not Parliament that should rule; it is the people who should rule through Parliament.

⚜ If you have a motor-car . . . you have to have a brake. A brake, in its essence, is one-sided; it prevents an accident through going too fast. It was not intended to prevent accidents through going too slow. For that you must look elsewhere. . . . You must look to the engine and of course to the petrol supply. For that there is the renewed impulse of the people's will; but it is by the force of the engine, occasionally regulated by the brake, that the steady progress of the nation and of society is maintained.

LEISURE

⚜ We have had a leisured class. It has vanished. Now we must think of the leisured masses.

⚜ By material well-being, I mean not only abundance but a degree of leisure for the masses such as has never been before in our mortal struggle for life.

LIBERALISM

✤ Liberalism has its own history and its own tradition. Socialism has its own formulas and aims. Socialism seeks to pull down wealth; Liberalism seeks to raise up poverty. Socialism would destroy private interest; Liberalism would preserve private interests in the only way they can be safely and justly preserved, namely by reconciling them with public right. Socialism would kill enterprise; Liberalism would rescue enterprise from the trammels of privilege and preference. Socialism assails the preeminence of the individual; Liberalism seeks, and shall seek more in the future, to build a minimum standard for the mass. Socialism exalts the rule; Liberalism exalts the man. Socialism attacks capital; Liberalism attacks monopoly.[37]

LIBERTY

✤ We have exalted liberty; it remains to preserve her.

LIBRARIES

✤ Nothing makes a man more reverent than a library.

LIES

✤ If truth is many-sided, mendacity is many-tongued.

✤ There are a terrible lot of lies going around the world, and the worst of it is half of them are true.

---

[37] Churchill as a Liberal made this attack on socialism in 1908.

LIFE

❧ What is the use of living, if it be not to strive for noble causes and to make this muddled world a better place to live in after we are gone?

❧ It has been a grand journey—well-worth making once.[38]

❧ Life is sensation; sensation is life.

❧ Life is a riddle: we shall learn the answer when we die.

❧ The human story does not always unfold like a mathematical calculation. . . . [T]he element of the unexpected and the unforeseeable is what gives some of its relish and saves us from falling into the mechanical thralldom of the logicians.

❧ There is no purpose in living where there is nothing to do.

LOGIC

❧ Logic is a poor guide compared to custom.

❧ Logic, like science, must be the servant and not the master of man.

LONDON

❧ Here in this strong City of Refuge which enshrines the little deeds of human progress.

---

[38] Possibly his last recorded statement—said in January 1965.

LOYALTY

‡ I have signed on for the voyage and would stick to the ship.[39]

‡ To change your mind is one thing; to turn on those who have followed your previous advice is another.

LUCK

‡ Life is a whole and luck is a whole and no part of them can be separated from the rest.

LUXURIES

‡ We shall forgo our luxuries but not our pleasures.

MAGNANIMITY

‡ Great countries . . . must not allow resentment or caprice or imitation or vengeance to enter into their policy.

‡ In war, resolution; in defeat, defiance; in victory, magnanimity; in peace, good will.[40]

‡ My hatred died with their surrender.[41]

MALAISE

‡ But what is the disease we are suffering from now in this island?

---

[39] To Harold Macmillan about an attempt to oust Chamberlain when Churchill was a member of the Chamberlain cabinet.

[40] Churchill chose these words for a tablet in a French town marker honoring the soldiers of World War I. He later adopted them as the motto for his World War II history, *The Second World War*.

[41] Churchill is referring to the Germans after their defeat in World War II.

It is a disease more dangerous than malaria. It is a disease of the will power.

MANKIND

⚜ Is it the only lesson of history that mankind is unteachable?

MARRIAGE

⚜ At Blenheim I took two very important decisions: to be born and marry. I am content with the decision I took on both occasions. I have never had cause to regret either.

MATERIALISM

⚜ No material progress, even though it takes shapes we cannot now conceive, or however it may expand the faculties of man, can bring comfort to his soul.

⚜ Material progress, however magnificent, will never accomplish the mission of mankind.

MATHEMATICS

⚜ [An] "Alice in Wonderland" world, at the portals of which stood "A Quadratic Equation" followed by the dim chambers inhabited by the Differential Calculus and then a strange corridor of Sines, Cosines, and Tangents in a highly square-rooted condition.[42]

MEDIA

⚜ How strange it is that the past is so little understood! We live in the most thoughtless of ages. Everyday headlines and short term views.

---

[42] About being Chancellor of the Exchequer in 1924.

MEDICINE

꽃 The only way to swallow a bitter mixture is to take it in a single gulp.

꽃 There ought to be a hagiology of medical science and we ought to have saints' days to commemorate the great discoveries which have been made for all mankind . . . a holiday, a day of jubilation when we can fete St. Anesthesia, and pure and chaste St. Antiseptic . . . and if I had a vote, I should be bound to celebrate St. Penicillin.[43]

MEDIOCRITY

꽃 They are a class of right honorable gentlemen—all good men—all honest men—who are ready to make great sacrifices for their opinions, but they have no opinions. They are ready to die for the truth if they only knew what the truth was.

꽃 I have no patience with people who are always raising difficulties.

꽃 We cannot have a band of drones in our midst, whether they come from the ancient aristocracy, the modern plutocracy, or the ordinary type of pub-crawler.

꽃 We live in an age of great events and little men.

꽃 When a man cannot distinguish a great from a small event, he is of no use.

---

[43] Churchill's life was probably saved in 1943 by penicillin when he was suffering from pneumonia.

MEMORIAL

✤ Honor to the brave who fell. Their sacrifice was not in vain.

MERCY

✤ One should be just before one is generous.

METTLE

✤ We have not journeyed across the countries, across the oceans, across the mountains, across the prairies because we are made of sugar candy.[44]

MIDDLE AGE

✤ We are happier in many ways when we are old than when we are young. The young sow wild oats, the old grow sage.

MILITARISM

✤ The statesman who yields to war fever must realize that once the signal is given he is no longer master of his policy.

MILITARY COOPERATION

✤ Why, you may take the most gallant sailor, the most intrepid airman, or the most audacious soldier, put them at a table together— what do you get? The sum total of their fears.

MISTAKES

✤ I am sure that the mistake of that time will not be repeated; we should probably make another set of mistakes.

---

44 Speech to Canadian House of Commons in December 1941.

MODESTY

❖ I am not usually accused even by my friends of a modest or retiring disposition.

MONARCHY

❖ A barrier against dictatorships.

❖ When our kings are in conflict with our constitution, we change our kings.

MONEY

❖ How it melts!

MOTHER

❖ My mother always seemed to me a fairy princess; a radiant being possessed of limitless riches and power. She shone for me like the Evening Star. I loved her dearly but at a distance.

MOTIVES

❖ Virtuous motives, trammeled by inertia and timidity, are no match for armed and resolute wickedness.

MOUNTAINS

❖ Freedom dwells among the mountains.[45]

---

[45] Churchill speaking in Wales in 1948.

MUNICIPAL UNIONS
✧ I decline to be utterly impartial between the Fire Brigade and the fire.

MUSTARD
✧ A gentleman does not have a ham sandwich without mustard.

NAPS
✧ A man should sleep sometime between lunch and dinner in order to be at his best in the evening when he joins his wife and friends at dinner.

✧ No half-way measures. Take off your clothes and get into bed.

NATIONALISM
✧ Who is the man vain enough to suppose that the long antagonisms of history and of time can in all circumstances be adjusted and compacted by the smooth and superficial conventions of politicians and ambitions?

✧ No race, country or individual has a monopoly of good or evil.

NATIONS
✧ The strength and character of a national civilization is not built up like a scaffolding or fitted together like a machine. Its growth is more like that of a plant or a tree. . . . No one should ever cut one down without planting another. It is very much easier to cut down trees than to grow them.

❧ A nation without a conscience is a nation without a soul. A nation without a soul is a nation that cannot live.

❧ Unless the intellect of a nation keeps abreast of all national components, the society in which that occurs is no longer progressing.

NATIONALITIES
❧ In dealing with nationalities nothing is more fatal than a dodge. Wrongs will be forgiven, sufferings and losses will be forgiven or forgotten, battles will be remembered only as they recall the martial virtues of the combatants; but anything like chicanery, anything like a trick will always rankle.

NATURE
❧ Nature is merciful and does not try her children, man or beast, beyond their compass.

THE NAVY
❧ Guard them well, admirals and captains, hardy tars and tall marines, guard them well and guard them true.[46]

❧ A few moments of exhilarating action to break the monotony of an endless succession of anxious uneventful days.[47]

❧ Dire crises might at any moment flash upon the scene with brilliant fortune or glare with mortal tragedy.

---

[46] Said as First Lord of the Admiralty in World War I.
[47] Churchill was describing the threat of submarine attack.

❧ To grudge the sailors a modern ship is as bad as grudging the safety-lamp to the miners.

NAZISM
❧ As Fascism sprang from Communism, so Nazism developed from Fascism.

❧ The foulest and most soul-destroying tyranny which has ever darkened and stained the pages of history.

❧ A boa constrictor that befouls its victims with saliva before engorging them.

NEGATIVISM
❧ People don't want anything done in any direction—fed-uppism.[48]

NEGOTIATION
❧ It would be an unmeasured crime to prolong this war for one unnecessary day. It would be an unmeasured and immeasurable blunder to make peace before the vital objects are achieved.

❧ I do not hold that we should rearm in order to fight. I hold that we should rearm in order to parley.

❧ Do not disband your army until you have got your terms.

❧ Let us have it on the basis that we are negotiating from strength and not from diversion.

---

[48] Churchill was speaking about Britain in the 1920s.

NEW YEAR'S DAY

✣ Here's to a year of toil—a year of struggle and peril, and a long step forward towards victory. May we all come through safe and with honour.[49]

NUCLEAR ENERGY

✣ We and all nations stand at this hour of human history upon the portals of supreme catastrophe and of immeasurable reward.[50]

NUCLEAR MISSILE

✣ Death stands at attention, obedient, expectant, ready to serve, ready to shear away the peoples *en masse;* ready if called on to pulverize without hope of repair what is left of civilization.[51]

✣ Might not a bomb no bigger than an orange be found to possess a secret power . . . to blast a township at a stroke . . . guided automatically in flying machines by wireless or other rays without a human pilot?[52]

NUCLEAR STALEMATE

✣ Let me tell you why in my opinion . . . a third World War is unlikely to happen. . . . Both sides know that it would begin with horrors of a kind and on a scale never dreamed of before by human beings.[53]

---

[49] Said on the train back from Ottawa on January 1, 1942.
[50] Said about the H-bomb in 1953.
[51] Churchill was predicting the horrors of the nuclear age in 1923.
[52] The prescient Churchill issued this prediction in a journal piece in 1924.
[53] Churchill in 1952 doubted World War III would ever take place.

✤ It may be that we shall by a process of sublime irony have reached a stage in this story where safety will be the sturdy child of terror, and survival the twin brother of annihilation.

✤ It is more likely to bring war to an end than an end to mankind.

✤ Certain mathematical quantities when they pass through infinity change their signs from plus to minus—It may be that this rule may have a moral application and that when the advance of destructive weapons enables everyone to kill everybody nobody will want to kill anyone at all.

NUCLEAR WAR
✤ The Dark Ages may return—the Stone Age may return on the gleaming wings of science; and what might now shower immeasurable material blessings upon mankind may even bring about its total destruction. Beware, I say! Time may be short.

✤ The fearful question confronts us: Have our problems got beyond our control?

OFFICE
✤ Far more important than the pleasing baubles of honor is the substantial gift of power.

OLD AGE
✤ It is difficult to find new interests at the end of one's life.

OPPONENTS
✤ I like a man who grins when he fights.

OPPORTUNITY

❖ We have before us a great opportunity, a golden opportunity, glittering bright but short-lived. Our chance is now at hand. The chance is there; the cause is there, the man is there.

❖ Remember the story of the Spanish prisoner. For many years he was confined in a dungeon. . . . One day it occurred to him to push the door of his cell. It was open; and it had never been locked.

OPPOSITION

❖ If I could not be conductor of the orchestra, I would like to be the kettle drum.

OPPOSITION PARTY

❖ The Opposition [is] not responsible for proposing integrated and complicated measures of policy. Sometimes [they] do but it is not [their] obligation.

ORATORS

❖ His ideas began to take the form of words to group themselves into sentences; he murmured to himself; the rhythm of his own language swayed him: instinctively he alliterated.

❖ Before he can inspire them with any emotion, he must be swayed by it himself. When he would rouse their indignation, his heart is filled with anger. Before he can move their tears, his own must flow. To convince them, he must himself believe. His opinions may change as their impressions fade but every orator means what he says at that moment he says it. He may be often inconsistent. He is never consciously insincere.

ORATORY

⚜ It was the nation and race dwelling all around the globe that had the lion's heart. I had the luck to be called upon to give the roar.[54]

⚜ I suppress with difficulty an impulse to become sententious.

⚜ Of all the talents bestowed upon men, none is so precious as the gift of oratory. He who enjoys it wields a power more durable than that of a great king.

⚜ It was my ambition all my life to be master of the spoken word.

⚜ The sentences of the orator when he appeals to his art become long, rolling and sonorous. The peculiar balance of the phrases produces a cadence which resembles blank verse rather than prose.

ORPHANS

⚜ What greater tragedy can there be than is presented by the spectacle of a child whose life prospects and hopes are smashed at the very outset of its existence?

OUTSPOKENNESS

⚜ It is human nature to dislike the wretch who dispels the delightful mirages and speaks roughly to the dupes.

---

54 Churchill's reply on his eightieth birthday celebration in 1954 that he was "the lion."

✢ I have in my life concentrated more on self-expression than on self-denial.

✢ There is no such thing as a negative virtue. If I have been of any service to my fellow men, it has never been by self-repression, but always by self-expression.

✢ All the years that I have been in the House of Commons I have always said to myself one thing: "Do not interrupt" and I have never been able to keep to that resolution.

OVEREXTENSION

✢ He created a spider's web but forgot the spider.[55]

✢ To try to be safe everywhere is to be strong nowhere.

PACIFISM

✢ I have always been against the Pacifists during the quarrel and against the Jingoists at its close.

PAIN

✢ By a blessed dispensation, human beings forget physical pain much more quickly than they do their joyous emotions and experiences. A merciful Providence passes the sponge of oblivion across much that is suffered, and enables us to cherish the great moments of life and honor which come to us in the march.

---

[55] Hitler's mistake in defending conquests in the Balkans and depriving his armies in France of reinforcements.

PAINTING

❧ Painting is a friend, who makes no undue demands, excites to no exhausting pursuits, keeps faithful pace even with feeble steps and holds her canvas as a screen between us and the envious eyes of time or the surly advance of Decrepitude.

❧ Anyone could see that it [the canvas] could not hit back. No evil fate avenged the jaunty violence. The canvas grinned in helplessness before me. The spell was broken. The sickly inhibitions rolled away. I seized the largest brush and fell upon my victim with berserk fury. I have never felt any awe of a canvas since.

❧ I cannot pretend to feel impartial about colors. I rejoice with the brilliant ones and am genuinely sorry for the poor browns. When I get to heaven I mean to spend a considerable portion of my first million years in painting and so get to the bottom of the subject. But then I shall require a still gayer palette than I get here below. I expect orange and vermilion will be the darkest and dullest colors and beyond them there will be a whole range of wonderful new colors which will delight the celestial eye.

❧ I prefer landscapes. A tree doesn't complain that I haven't done it justice.

❧ Look around these walls! We see reflections from hours of pleasure, hours of intense creative enjoyment, bottled sunshine, captured inspiration, perennial delight.

PANIC

❧ It is very much better sometimes to have a panic feeling beforehand, and then be quite calm when things happen, than to be extremely calm beforehand and to get into a panic when things happen.

PARTY LOYALTY

✣ A change of Party is usually considered a much more serious breach of consistency than a change of view.

✣ Some men change their Party for the sake of their principles; others change their principles for the sake of their Party.

✣ Anyone can rat—it takes a bit of ingenuity to re-rat.[56]

PAST

✣ If we open a quarrel between the past and the present, we shall find that we have lost the future.

PATIENCE

✣ I am certainly not one who needs to be prodded. In fact, if anything I am a prod. My difficulties rather lie in finding the patience and self-restraint to wait through many anxious weeks for the results to be achieved.

PEACE

✣ Why should war be the only cause large enough to call forth great and free countries?

---

[56] On his decision to leave the Liberal party and become a Conservative again in 1922. Originally a Conservative, Churchill switched over to the Liberal party in 1904.

✤ Peace will not be preserved by pious sentiments. It will not be preserved by casting aside in dangerous times the panoply of warlike strength.

✤ How many wars have been averted by patience and good will?

✤ We have now reached a place in the journey where . . . it must be world anarchy or world order.

PERFECTIONISM
✤ The maxim, "Nothing avails but perfection," spells paralysis.

PERSISTENCE
✤ We shall come through! We cannot tell when, we cannot tell how, but we shall come through.

PETTINESS
✤ Great events and personalities are all made small when passed through the medium of a small mind.

PLANNERS
✤ Those whose minds are attracted or compelled to rigid and symmetrical systems of government should remember that logic, like science, must be the servant and not the master of man. Human beings and human societies are not structures that are built or machines that are forged. They are plants that grow and must be treated as such.

PLANNING

✢ It is better to have an ambitious plan than none at all.

✢ Large views always triumph over small ideas.

✢ Any clever person can make plans for winning a war if he has no responsibility for carrying them out.

✢ It is one thing to see the forward path and another to be able to take it.

✢ One of my fundamental ideas has always been the importance of keeping as many options as possible open to serve the main purpose, especially in time of war.

✢ One [anecdote] is the celebrated tale of the man who gave the powder to the bear. He mixed the powder with the greatest care, making sure that not only the ingredients but the proportions were absolutely correct. He rolled it up in a large paper spill, and was about to blow it down the bear's throat. But the bear blew first.

✢ Our hope is to regulate the unthinkable.

PLATITUDES

✢ This is no time for windy platitudes and glittering advertisements. . . . [We] had far better go down telling the truth and acting in accordance with the virtues of our position than gain a shabbily-bought office by easy and fickle froth and chatter.

✢ There is no greater mistake than to suppose that platitudes, smooth words, and timid policies offer a path to safety.

PLEASURE
✤ Every pleasure has its corresponding drawback, just as every rose its thorn.

POLAND
✤ There are few virtues which the Poles do not possess and there are few errors they have ever avoided.

✤ Poland must be mistress in her own house, and captain of her own soul.

✤ The soul of Poland is indestructible. . . . [S]he will rise again like a rock, which may for a spell be submerged by a tidal wave, but which remains a rock.

POLICY
✤ We must strive to combine the virtues of wisdom and of daring.

✤ We have to select from a host of dangers the one which can best be dealt with and which, if dealt with, causes others to fall away.

✤ In my experience of large enterprises, I have found it is often a mistake to try to settle everything at once.

POLITICIANS
✤ The world today is ruled by harassed politicians absorbed in getting into office or turning out the other man so that not much room is left for debating great issues on their merits.

✦ The main qualification for political office is the ability to foretell what is going to happen tomorrow, next week, next month and next year. And to have the ability afterwards to explain why it didn't happen.

✦ Politicians know they are but the creatures of the day. They hold no golden casket enshrining the treasures of centuries to be shattered irretrievably in their hands.

✦ Politicians rise by toil, not struggles. They expect to fall; they hope to rise again.

✦ A bad politician is one you disagree with.

✦ I have always felt that a politician is to be judged by the animosities he excites among his opponents.

POLITICS
✦ It is a very serious thing for a political creed or a political party when they are compelled in spite of themselves to hail national misfortunes as a means of advancing their cause.

✦ It would be a great reform in politics if wisdom could be made to spread as easily and as rapidly as folly.

✦ It is a fine game to play the game of politics—and it's well worth waiting for a good hand before really plunging.

✦ Politics is almost as exciting as war and quite as dangerous. In war you can only be killed once, but in politics many times.

POLLS

✢ Nothing is more dangerous . . . than to live in the temperamental atmosphere of a Gallup Poll—always feeling one's pulse and taking one's temperature.

POWER

✢ When one has reached the summit of power and surmounted so many obstacles, there is a danger of becoming convinced that one can do anything one likes and that any strong personal view is necessarily acceptable to the nation and can be enforced upon one's subordinates.

PRAGMATISM

✢ We do not think logic and clear-cut principles are necessarily the sole keys to what ought to be done in swiftly changing and indefinable situations. . . . [W]e assign a larger importance to opportunism and improvisation, seeking rather to live and conquer in accordance with the unfolding event than to aspire to dominate it often by fundamental decisions.

✢ My conscience is a good girl. I can come to terms with her.

✢ Never stand so high upon a principle that you cannot lower it to suit the circumstances.

PREJUDICE

✢ Houses are built of bricks, mortar and good will, not politics, prejudices and spite.

❧ All men are equal and all distinctions between them are unhealthy and undemocratic.

PREPAREDNESS
❧ The first victory we have to win is to avoid a battle; the second if we cannot avoid it, to win it.

❧ We must be prepared. . . . [I]t is good to be patient, it is good to be circumspect, it is good to be peace-loving—but it is not enough. We must be strong, we must be self-reliant.

PRESS
❧ The essence of American journalism is vulgarity divested of truth.

❧ The naggers in the press are not without resource.

PRIME MINISTER
❧ The dignity of a Prime Minister, like a lady's virtue, is not susceptible of partial diminution.

❧ The high belief in the perfection of man is appropriate in a man of the cloth but not in a prime minister.

PRINCIPLES
❧ It is always more easy to discover and proclaim general principles than to apply them.

❧ Do principles change with dates?

❧ If a government has no moral scruples it often seems to gain great advantages and liberties of action, but "all comes out even at the end of the day, and all will come out yet more even when all the days are ended."

PRIVATE ENTERPRISE
❧ We are for private enterprise with all its ingenuity, thrift and contrivance, and we believe it can flourish best within a strict and well-understood system of prevention and correction of abuses. In a complex community like our own no absolute rigid uniformity of practice is possible.

❧ Some see private enterprise as a predatory animal to be shot, others look on it as a cow to be milked but a few see it as a sturdy horse pulling a wagon.

❧ What is the characteristic that all craftsmen, pioneers and individualists have in common? Surely it is the individual effort of hard work and brains and the development of free and independent enterprise.

❧ If you destroy a free market, you create a black market.

PRIVATE PROPERTY
❧ Private property has a right to be defended. Our civilization is built up by private property and can only be defended by private property.

✤ What we mean is a *personal* property-owning democracy. Households which have possessions which they prize and cherish because they are their own, or even a house and garden of their own, a little money put by for a rainy day, or an insurance policy—that is what the Conservatives mean by a property-owning democracy.

PROBLEMS

✤ Out of intense complexities, intense simplicities emerge.

✤ In critical and baffling situations, it is always best to return to first principle and simple action.

✤ The present problem cannot be cured by anything slick, cheap, swift and impatient. It is to be done by hard work in many spheres of action.

✤ The question which we must ask ourselves is not whether we like or do not like what is going on, but what we are going to do about it.

✤ In my experience of large enterprises, I have found it is often a mistake to try to settle everything at once.

✤ Don't argue the matter. The difficulties will argue for themselves.

PROCRASTINATION

✤ The era of procrastination, of half-measures, of soothing and baffling expedients, of delays is coming to its close.

PROMISES

⚜ You must never make a promise which you do not fulfill.

PROPHECY

⚜ I always avoid prophesying beforehand, because it is much better policy to prophesy after the event has already taken place.

⚜ There are two processes . . . when we try to prophesy. We can seek a period in the past whose conditions resemble as closely as possible those of our day. . . . Secondly, we can survey the general course of development in our immediate past and endeavor to prolong it into the near future. The first is the method of the historian; the second that of the scientist.

PROPHETS

⚜ A hopeful disposition is not the sole qualification to be a prophet.

⚜ Every prophet has to come from civilization, but every prophet has to go into the wilderness. He must have a strong impression of a complex variety and all that it has to give and he must serve a period of isolation and meditation. This is the process by which psychic dynamite is made.

PSYCHIATRY

⚜ Some of you doctors get queer ideas about what's in people's heads. You think too much about these things. It isn't healthy.

PUNCTUALITY

❧ Unpunctuality is a vile habit, and all my life I have tried to break myself of it.

❧ I am a sporting man. I always like to give trains and airplanes a fair chance of getting away.

PURPOSE

❧ A cavalry charge is very like ordinary life. So long as you are all right, firmly in your saddle, your horse in hand and well armed, lots of enemies will give you wide berth.

❧ A weakening in our purpose and therefore in our unity—that is the mortal crime.

❧ Why is it the ship beats the waves when the waves are so many and the ship is one? The reason is that ship has a purpose.

QUOTATIONS

❧ It is a good thing for an uneducated man to read books of quotation. . . . The quotations when engraved upon the memory give you good thoughts.

❧ Verify your quotations.

RACISM

❧ Duty and prudence alike command . . . that the germ centers of hatred and revenge shall be constantly and vigilantly curbed and bested in good time and that adequate organization should be set up to make sure that the pestilence can be controlled at its earliest beginning before it spreads and rages throughout the entire earth.

RALLYING
✣ Look forward, do not look backward. Gather afresh in heart and spirit all the energies of your being, bend anew together for a supreme effort.

RATIONING
✣ In wartime, rationing is the alternative to famine. In peace, it may well become the alternative to abundance.

REALPOLITIK
✣ *Realpolitik* meant that the standards of morality in international affairs could be ignored whenever material advantage might be gained.

✣ The only real guide to the actions of mighty nations and powerful governments is a correct estimate of what are and what they consider to be their own interests.

THE RED CROSS
✣ The Red Cross was a platform, the only platform between lines of battle; men could meet there together and recognize their common humanity, and the value of an international code of law and convention.

REFORM

✧ If the train is running on the wrong line downhill at sixty miles an hour, it is no good trying to stop it by building a brick wall across the track. That would only mean the wall was shattered, the train was wrecked, and the passengers maimed. First, you have to put on the brakes . . . then the engine has to be put in reverse.

RESIGNATION

✧ To resign is not to retire.

RESOLUTION

✧ Never take "No" for an answer. Never submit to failure.

✧ Never give in! Never give in! Never, never, never, never—in nothing great and small—large and petty—Never give in except to convictions of honour and good sense.[57]

RESPONSIBILITY

✧ Perhaps it is better to be irresponsible and right than responsible and wrong.

✧ It is no use saying "We are doing our best." You have to succeed in doing what is necessary.

✧ The price of greatness is responsibility.

✧ We cannot escape our dangers by recoiling from them.

---

[57] Spoken to Harrow School in 1941.

✢ A man who says "I disclaim responsibility for failure" cannot be the final arbiter of the measures which may be found vital to success.

RETICENCE
✢ Too often the strong silent man is silent because he doesn't know what to say.

RETIREMENT
✢ Tranquility! There is nothing more tranquil than the grave. I will never stifle myself in such a moral and intellectual sepulchre.

✢ It would be easy for me to retire in an odor of civic freedom.[58]

✢ If I stay, for the time being, bearing the burden at my age, it is not because of love for power or office. I have had an ample share of both. If I stay it is because I have the feeling that I may, through things that have happened, have an influence on what I care about above all else—the heralding of a sure and lasting peace.[59]

REVENGE
✢ Revenge is, of all satisfactions, the most costly and long drawn out; retroactive persecution is . . . the most pernicious.

---

[58] On the prospect of retiring after defeat in 1945.
[59] After a stroke in June, Prime Minister Churchill addressed the Conservative party convention in October 1953 signaling his intention not to resign.

RHETORIC

⚜ There is no worse mistake in public leadership than to hold out false hopes soon to be swept away.

RICH

⚜ Wealth, taste and leisure can bring many things but they do not bring happiness.

RIOTING

⚜ I say that the last thing that represents democracy is mob law.

RISK

⚜ Success cannot be guaranteed. There are no safe battles.

⚜ Victory will never be found by taking the line of least resistance.

RUSSIA

⚜ I cannot forecast to you the action of Russia. It is a riddle wrapped in a mystery inside an enigma.

⚜ I deem it highly important that we shake hands with the Russians as far east as possible.[60]

⚜ Their worst misfortune was Lenin's birth; their next worst was his death.

---

[60] Message to General Dwight David Eisenhower in 1945.

From what I have seen of our Russian friends and allies during the war I am convinced that there is nothing they admire so much as strength and nothing for which they have less respect than weakness—particularly military weakness.[61]

SACRIFICE
Out of the depths of sorrow and sacrifice will be born again the glory of mankind.

SANDWICHES
The bread must be wafer-thin. It is nothing more than a vehicle to convey the filling to the stomach.

SAVINGS
If you strike at savings you at once propagate the idea of "Let us eat, drink and be merry, for tomorrow we die."

SCHOOL
This interlude of school makes a somber gray patch upon the chart of my journey. It was an unending spell of worries that did not then seem petty, and of toil uncheered by fruition; a time of discomfort, restriction and purposeless monotony.

SCIENCE
Surely there never was an army which advanced like the army of science.

---

[61] Iron Curtain address in Fulton, Missouri, in 1945.

❧ The only limits to human progress are those made by our own shortcomings. Science is ready to extend the frontiers of every country without injury to the rights of others and to increase the well-being of every people at the expense of none.

❧ Humanity stands today at its most fateful milestone. On the one hand science opens up a chasm of self-destruction beyond limits. On the other hand she displays a vision of plenty and comfort of which the masses of no race have ever known or even dreamed.

❧ Man in this moment of his history has emerged in greater supremacy over the forces of nature than has ever been dreamed of before. He has it in his power to solve quite easily the problems of material existence. He has conquered the wild beasts and he has even conquered the insects and microbes. All is in his hand. He has to conquer his last and worst enemy—himself.

❧ Science, which now offers us a golden age with one hand, offers at the same time with the other the doom of all that we have built up inch by inch since the Stone Age and the dawn of any human annals. My faith is in the high progressive destiny of man. I do not believe we are to be flung back into abysmal darkness by those fiercesome discoveries which human genius has made. Let us make sure that they are servants, but not our masters.

❧ Science burrows its insulated head in the filth of slaughterous inventions.

SECRETS
❧ It is wonderful how well men can keep secrets they have not been told.

SELF-INTEREST

❧ How vain are human calculations of self-interest![62]

SHORTSIGHTEDNESS

❧ Want of foresight, unwillingness to act . . . lack of clear thinking, confusion of counsel . . . these are the features which constitute the endless repetition of history.

SINCERITY

❧ Before he [the orator] can inspire them with any emotion, he must be swayed by it himself. When he would rouse their indignation, his heart is filled with anger. Before he can move their tears, his own must flow. To convince them, he must himself believe. His opinions may change as their impressions fade, but every orator means what he says at the moment he says it. He may be often inconsistent. He is never consciously insincere.

SOCIALISM

❧ Socialism assails the pre-eminence of the individual.

❧ Whenever socialism has been tried, it has failed.

❧ If I were asked the difference between Socialism and Communism, I could only reply that the Socialist tries to lead us to disaster by foolish words and the Communist could try to drive us there by violent deeds.

---

[62] Churchill was referring to French Admiral Darlan's self-serving pro-German policies in North Africa in 1942.

✣ Socialism is inseparably interwoven with totalitarianism and the abject worship of the State. . . . This state is to be the arch-employer, the arch-planner, the arch-administrator and ruler, and the arch-caucus-boss.

✣ Government of the duds, by the duds and for the duds.

✣ The Socialist dream is no longer Utopia but Queue-topia!

✣ No Socialist system can be established without a political police.

✣ It is not alone that property, in all its forms, is struck at, but that liberty, in all its forms, is challenged by the fundamental conceptions of Socialism.

✣ I am told of the popular slogan "Labour [Socialist Government] gets things done" . . . but surely it should run "Labour gets things done in."

✣ Socialism is the philosophy of failure, the creed of ignorance and the gospel of envy.

✣ I do not wonder that British youth is in revolt against the morbid doctrine that nothing matters but the equal sharing of miseries; that what used to be called the submerged tenth can only be rescued by bringing the other nine-tenths down to their level; against the folly that it is better that everyone should have half rations rather than that any by their exertions, or ability, should earn a second helping.

✣ Socialism is contrary to human nature.

"All men are created equal" says the American Declaration of Independence. "All men shall be kept equal" say the Socialists.

SOCIAL SECURITY

If I had my way, I would write the word "insure" over the door of every cottage and upon the blotting book of every public man.

I am glad to have a hand in all that structure of pensions and insurance which no other country can rival.

SOLDIERING

The more I see of soldiering, the more I like it, but the more I feel convinced that it is not my metier.

The course of the soldier is not really a contempt for physical evils and indifference to danger. It is a more or less successful attempt to simulate these habits of mime. Most men aspire to be good actors in a play. . . . Three principal influences combine to assert men in this attempt: preparation, vanity and sentiment.

SOLICITATION

Give us the tools and we will finish the job.[63]

SOLUTIONS

To gain one's way is no escape from the responsibility of an inferior solution.

---

[63] Churchill asking for American military aid in 1941.

✤ You cannot cure cancer by a majority. What is wanted is a remedy.

✤ Experience shows that the only safe plan for human action is to act with great simplicity and give judgment on the merits of questions at each particular stage.

SOLVENCY
✤ All social reform . . . which is not founded upon a stable medium of internal exchange becomes a swindle and a fraud.

✤ Solvency is valueless without security and security is impossible to achieve without solvency.

SOVIET UNION
✤ I do not believe that Soviet Russia desires war. What they want are the fruits of war.

SPECIALIZATION
✤ We want a lot of engineers in the world but we do not want a world of engineers.

SPEECH
✤ A series of facts . . . brought forward all pointing in a common direction. The end appears in view before it is reached. The word anticipates the conclusion and the last words fall amid a thunder of assent.

✤ I am going to give a long speech today. I haven't had time to prepare a short one.

✢ Don't deliver an essay with so many points. No one can absorb it. Just say one thing. . . . Of course, you can say the point in many different ways over and over again with different illustrations.

✢ Vary the pose and vary the pitch. Finally, don't forget the pause.

SPENDING
✢ Money may be used as a sop, or it may be used as a lever. When you use it as a sop, you get your cheer, you get some friendly notes in the Press. . . . But it is a sop and it is gone. But if you use it as a lever it may be made to influence matters of far greater consequence than is measured even by the actual amount involved.

SPORTS
✢ In sport, in courage, and in the sight of Heaven, all men must meet on equal terms.

STANDARDS
✢ Let us set up a standard around which the brave and loyal can rally.

STATESMANSHIP
✢ A statesman in contact with the moving current of events and anxious to keep the ship on an even keel and steer a steady course may lean all his weight now on one side and now on the other. His arguments in each case when contrasted can be shown to be not only very different in character, but contradictory in spirit and opposite in direction; yet his object will throughout have remained the same.

✤ Statesmen are not called upon to settle the easy questions. These often settle themselves. It is when the balance quivers and the proportions are veiled in mist that the opportunities for world-saving decisions present themselves.

STATISTICS
✤ You cannot ask us to take sides against arithmetic.

STRATEGY
✤ First there is the period of consolidation, [then] of combination, and [then] of final preparation.

✤ The best method of acquiring flexibility is to have three or four plans for all the probable contingencies all worded out with the utmost detail. Then it is much easier to switch from one to the other as to when and where the cat jumps.

✤ Any clever person can make plans for winning a war if he has no responsibility for carrying them out.

STRENGTH
✤ A great principle only carries weight when it is associated with the movement of great forces.

✤ Peace is our aim and strength the only way of getting it.

STUDENTS
✤ Take full advantage of these years when the wisdom of the world is placed at your disposal, but do not spend too much time

buckling on your arms in the tent. The battle is going on in every walk and sphere of life.

STUPIDITY
❧ It would be a great reform in politics if wisdom could be made to spread as easily and rapidly as folly.

❧ Unwisdom prevailed.

SUFFERING
❧ We shall draw from the heart of suffering itself the means of inspiration and survival.

SUPPLIES
❧ Victory is the beautiful bright-colored flower. Transport is the stem without which it could never have blossomed. Yet even the military student, in his zeal to master the fascinating combinations of the actual conflict, often forgets the far more interesting complications of supply.

SURRENDER
❧ Nations which went down fighting rose again, but those which surrendered tamely were finished.

TANKS
❧ The tank was originally invented to clear a way for the infantry in the teeth of machine gun fire. Now it is the infantry who will have to clear a way for the tanks.

TASTES

✣ My tastes are simple; I am easily satisfied with the best.

TAXATION

✣ Thrift is penalized by the heaviest taxation. . . . Regulations increasingly take the place of statutes.

✣ I am not in favor of the present rate of taxation on earned income. It will destroy all incentive.

✣ The idea that a nation can tax itself into prosperity is one of the cruelest delusions which has ever befuddled the human mind.

TAXES

✣ Whoever heard of a country trying to get rich by putting in taxes.

✣ I was brought up to believe that taxation was a bad thing but the consuming power of the people was a good thing.

✣ A grave discouragement to enterprise and thrift.

TECHNOLOGY

✣ Unless the intellect of a nation keeps abreast of all material improvements, the society in which that occurs is no longer progressing.

TIME

✣ Time is neutral; but it can be made the ally of those who will seize it and use it to the full.

TIMIDITY

✣ I draw a distinction between mistakes. There is the mistake which comes through daring—what I call a mistake towards the enemy—in which you must sustain *your* commanders. . . . There are mistakes from the safety-first principle—mistakes of turning away from the enemy; and they require a far more acid consideration.

TOBACCO

✣ I never smoke before breakfast.

✣ Tobacco is bad for love; but old age is worse.

✣ How can I tell that my temper would have been as sweet or my companionship as agreeable if I had abjured from in my youth the goddess Nicotine?

TOTALITARIANISM

✣ A one-man state is not a state.

✣ I discharge a duty to the human species in breaking down a military despotism. I do not like to see a government supported only by bayonets.

✣ Totalitarianism is only state-organized barbarism.

✣ I have never taken the view that individuals exist to serve a state or system.

✣ There are two non-God religions. Nazism and Communism—two peas . . . Tweedledum and Tweedledee. You leave out God and you substitute the Devil. You leave out love and you substitute hate.

TRADITION

✢ Fortune is rightly malignant to those who break with the customs of the past.

TRADE

✢ I am very much in favor of "trade not aid."

TRANSATLANTIC VOYAGES

✢ On the first day out your stomach is so upset you keep to your staterooms. The second day you at last begin to gain your sea legs and venture tentatively to the deck. By the third day when you're really fit and ready for romance, you're about to dock.

TRUTH

✢ The truth is uncontrovertible. Panic may resent it; ignorance may deride it; malice may destroy it, but there it is.

✢ Truth is so precious that she must often be attended by a bodyguard of lies.[64]

✢ We are told that we must not alarm the easygoing voter and public. How thin and paltry these arguments will sound if we are caught a year or two hence, fat, opulent and defenseless.

---

[64] Churchill said this at the Teheran Conference in 1943 about European invasion plans.

TYRANNY

✤ Tyranny is our foe, whatever trappings or disguise it wears, whatever language it speaks, be it external or internal, we must forever be on our guard, ever mobilized, ever vigilant, always ready to spring at its throat.

✤ I always have been opposed to tyranny in every guise. It makes no difference to me what dress it wears, what slogans it mouths.

✤ Tyrannies may restrain or regulate their words. The machinery of propaganda may pack their minds with falsehood and deny them truth for many generations of time. But the soul of man thus held in trance or frozen in a long night can be awakened by a spark coming from God.

✤ Governments who have seized upon power by violence and by usurpation have often resorted to terrorism in their desperate efforts to keep what they have stolen.

UNEMPLOYMENT

✤ We must try to seek the remedies of the disease, not merely the remedies for the symptoms.

UNITED NATIONS

✤ The U.N. was set up not to get us to Heaven but only to save us from Hell.

UNITED STATES CONSTITUTION

✤ The great men who founded the American Constitution embodied this separation of authority in the strongest and most durable form. Not only did they divide executive, legislative and judicial functions, but also by instituting a federal system they preserved immense and sovereign rights to local communities and by all these means they have preserved—often at some inconvenience—a system of law and liberty under which they have thrived and reached the leadership of the world.

❖ No constitution was ever written in better English.

UNITY
❖ National unity does not mean national unanimity.

❖ We are plunged in a long and grievous struggle. But all will come right if we all work together to the end.

UNIVERSITY
❖ The first duty of a university is to teach wisdom not a trade; character not technicalities.

❖ The university is the treasure house of a country's future.

❖ He who has received a university training possesses a rich choice. He need never be inactive or bored, there is not reason for him to seek refuge in the clack and clatter of our modern life. He need not be dependent on the headlines which give him something new every day. He has the wisdom of all time to drink from, to enjoy as long as he lives.

❖ Young people study at universities to achieve knowledge, and not to learn a trade. We must learn to support ourselves, but we must also learn how to live.

❧ The more the opportunities of university education in any country are used, the brighter and healthier will the life of that country become.

❧ The university education is the guide to the reading of a lifetime.

UNREST
❧ It might be that the time will come when we enter on an era of kaleidoscopic politics—and governments will rise and fall and succeed each other with baffling speed.

VACATION
❧ Sunshine is my quest.

VALOR
❧ What a glory shines on the brave and true!

VANITY
❧ The vice that promotes so many virtues.

VEGETARIANS
❧ I will have meat; carnivores will win this war.

VERBOSITY
❧ It is sheer laziness not compressing thought into a reasonable space.

VERITIES

❧ All the greatest things are simple and can be expressed in a single word: Freedom, Justice, Honor, Duty, Mercy, Hope.

VICES

❧ Never trust a man who has not a single redeeming vice.

❧ My religion prescribed an absolute sacred rite of smoking cigars and drinking alcohol.

❧ If this is a world of vice and woe, I'll take the vice and you can have the woe.

VICTORY

❧ We were so glutted with victory that we cast it away.[65]

❧ I have always laid down the doctrine that the redress of grievances of the vanquished should precede the disarmament of the victors.

❧ The problems of victory are more agreeable than those of defeat, but they are no less difficult.

❧ When I speak of victory, I am not referring to those victories which crowd the daily placards of any newspaper. I am

---

[65] This was said about World War I in a speech to the House of Commons in 1940.

speaking of victory in the sense of a brilliant and formidable fact shaping the destinies of nations and shortening the direction of the war.

✤ The shadow of victory is disillusion: the reaction from extreme effort and prostration.

VISION
✤ It is always wise to look ahead, but difficult to look farther than you can see.

✤ It is a mistake to look too far ahead. Only one link in the chain of destiny can be handled at a time.

VOCABULARY
✤ I picked up a wide vocabulary and had a liking for words and for the feel of words fitting and falling into their places like pennies in a slot.

VOCATION
✤ A man's life must be nailed to the cross of either Thought or Action.

VOTERS
✤ At the bottom of all the tributes paid to democracy, is the little man walking into the booth, with a little pencil, making a little cross on a little bit of paper.

VOTING

❧ We must not forget what votes are. Votes are the means by which the poorest people in the country and all people in the country can make sure that they get their vital needs attended to.

❧ Apathy, complacency, illness, chatter or indifference may often be faults. On [election day] they will be crimes.

WAR

❧ Undertake no operation which is more costly to us in life than to the enemy.

❧ You can't run a war as if you were in a laboratory.

❧ The story of the human race is war. Except for brief and precarious interludes there has never been peace in the world.

❧ Never maltreat the enemy by halves.

❧ War, which used to be cruel and magnificent, has now become cruel and squalid.

❧ War is mainly a catalogue of blunders.

❧ War is not to be likened to a task which can be completed by installments, bit by bit, part of which when done is done forever. War is rather a race of an extraordinary character which, once started, has to be run through to the end.

❧ It is not possible in a major war to divide military from political affairs.

❧ War is horrible but slavery is worse.

❧ War attracts me and fascinates my mind with its tremendous situations. What vile and wicked folly and barbarism it all is.[66]

❧ [War] has at last been stripped of glitter and glamour. No more may Alexander, Caesar, and Napoleon lead armies to victory, ride their horses on the field of battle sharing the perils of their soldiers and deciding the fate of empires by the resolves and gestures of a few intense hours. For the future they will sit surrounded by clerks in offices, as safe, as quiet, and as dreary as government departments, while the fighting men in scores of thousands are slaughtered or stifled over the telephone by machinery.

❧ To some the game of war brings prizes, honor, advancement or experience; to some the consciousness of duty well discharged. . . . But here were those who had drawn the evil numbers—who had lost their all, to gain only a soldier's grave. Looking at those shapeless forms confined in a regulation blanket, the pride of race, the pomp of Empire, the glory of war appeared but the faint unsubstantial fabric of a dream.

❧ No compromise with the main purpose, no peace till victory, no pact with unrepentant wrong.[67]

❧ Never, never, never believe any war will be smooth and easy, or that anyone who embarks on the strange voyage can measure the tides

---

[66] The first sentence is sometimes quoted from a Churchill talk in 1903 as an indication that Churchill was a warmonger. Unless the succeeding sentence is added, it is out of context.
[67] Churchill delivered this reply to those who wanted a truce in 1917.

and hurricanes he will encounter. The statesman who yields to the war fever must realize that once the signal is given, he is no longer the master of policy.

⚜ War is very cruel. It goes for so long.

⚜ Wars are not won by evacuations.

WEAKNESS
⚜ Fatuity and fecklessness, which though devoid of guile, are not devoid of guilt.

⚜ Woe betide the political party or the public men who yield themselves to fatal promptings of lethargy or weakness.

⚜ Power is overwhelming until it is cast away. The shame is that our moral and intellectual guidance should not have been exerted as our material power. It is the contrast between the vague and soothing political sentiments on the one hand and the rough practical measures which have to be taken.

⚜ Virtuous motives, trammeled by inertia and timidity, are no match for armed and resolute wickedness.

WELFARE
⚜ We do not seek to pull down improvidently the structures of society but to erect balustrades upon the stairway of life, which will prevent helpless or foolish people from falling into the abyss.

⚜ The great security of a state depends upon the family systems, but that system cannot be properly maintained unless the hazards of life are safeguarded against.

WHISKEY

✷ When I was a young sub-altern in the South Africa campaign, the water was not fit to drink. To make it palatable, we had to add whiskey. By diligent effort I learned to like it.[68]

✷ Have some whiskey and soda water. It is a good drink to draw a sword on.

WIFE

✷ Twelve times I have seen your birthday come and each time your gracious beauty and loving charm have made a deeper impression on my heart. God bless you, darling, in the year that opens and give you happiness which fills your life.

✷ The most precious thing I have in life is your love for me.[69]

✷ I love you more each month that passes and feel the need of you and your beauty. . . . I am so devoured of egoism that I would like to have another soul in another world and meet you in another setting and pay you all the love and honor of the great romances.[70]

---

[68] Churchill's principal preference in whiskey was Johnnie Walker Red Label. He was a friend of Alexander Walker, the early head of the Scottish distillery.
[69] Written in 1920 to his wife.
[70] Written in a letter to his wife from the World War I trenches in France in 1916.

✢ Time passes swiftly but is it not joyous to see how great is the treasure we have gathered together?

WISDOM
✢ All wisdom is not new wisdom.

WORLD WAR II
✢ There never was a war in all history easier to prevent by timely action.

WORRY
✢ When I look back on all these worries, I remember the story of the old man who said . . . that he had a lot of issues in his life, most of which never happened.

✢ Worry is a spasm of the emotion.

WRITING
✢ Writing a book is an adventure. To begin with it is a toy, an amusement; then it becomes a mistress, and then a master and then a tyrant.

✢ Writing a long and substantial book is like having a friend and companion at your side to whom you can always turn for comfort and amusement and whose society becomes more attractive as a new and widening field of interest is lighted in the mind.

YOUTH

⚜ A young man cannot expect to get very far in life without some good smacks in the eye.

⚜ The world . . . was made to be won and wooed by youth.

⚜ Twenty to twenty-five. These are the years! Don't be content with things as they are.

⚜ No boy or girl should ever be disheartened by lack of success in their youth but should diligently and faithfully continue to persevere and make up for lost time.

ZIONISM

⚜ I am a Zionist. Let me make that clear. I was one of the original ones after the Balfour Declaration and I have worked faithfully at it.

Orations
and Perorations

Those who wish to savor the full majesty of Churchill's oratory cannot find it in a quoted sentence or two. The array of words, the cadence of line, the mounting of sentences, the crescendoing culmination can be sensed only by reading aloud the addresses as a whole. Of course, listening to his recorded addresses is even better.

Churchill has a rhetorical style. Besides his father, Lord Randolph, Bourke Cockran, the Irish-American politician, was the orator Churchill most emulated. Churchill's stately sweep of phrasing also reflects his reading of the historian Edward Gibbon.

Space requirements do not permit printing in full even his greatest addresses. Churchill addressed audiences for almost seven decades. Those talks fill eight volumes, with over a thousand pages to each volume.

Yet a sampling of his oratory can be offered through excerpts. This chapter presents excerpts of addresses from the would-be candidate to the retired statesman.

"RISING TIDE OF TORY DEMOCRACY"

1897, Primrose League, Bath, England

This was Churchill's first public speech. Churchill, when on military leave in May, had walked by the Conservative Party's office and found to his amazement that the headquarters was looking for volunteer speakers. Soon the twenty-two-year-old officer was asked to address the Primrose League in Bath. This was an auxiliary for the Conservative Party. It promoted allegiance to the crown and empire. It had been founded by Churchill's father, Lord Randolph. (The name Primrose was a reference to Benjamin Disraeli's favorite flower.)

For weeks Churchill prepared for his speaking debut. He wanted the speech to reflect the philosophy of his father, who had died two years earlier.

⚜ Conservative policy is a look-before-you-leap policy. . . . The British workingman has more to hope from the rising tide of Tory Democracy than from the dried-up drain pipe of Radicalism. . . . There were those who said that in this Jubilee year our Empire has reached the height of its glory and we should begin to decline. . . . Do not believe these croakers but give the lie to their dismal croaking by showing that the vigor and vitality of our race is unimpaired and that our determination is to uphold the Empire that we have inherited from our fathers as Englishmen, that our flags shall fly high on the sea, our voice be heard in the councils of Europe, our sovereign be supported by the love of her subjects. ⚜

THE BOER WAR (Maiden Address to Parliament)

Queen Victoria had died in February and Churchill had to interrupt an American speaking tour to return to the House of Commons. Churchill, a recent veteran of the Boer War, chose the British African colony situation as the subject for his maiden address. He would close this February speech to the House of Commons with a reference to his late father, who had served with many of the House of Commons members who were present.

✤ If the Boers remain deaf to the voice of reason and blind to the hands of friendships, if they refuse all overtures and disdain all terms . . . we can only hope that our own race . . . will show determination as strong and endurance as lasting as theirs. . . .

I cannot sit down without saying how grateful I am for the kindness and patience with which the House has heard me and which has been extended to me, I well know, not on my own account but because of a certain splendid memory which many honourable members still preserve. ✤

"A NET ON THE ABYSS"

As the doctrine of Socialism flourished, Churchill decided to attack the philosophy of collectivism. In an October speech to a Liberal

Party audience in Scotland, Churchill delivered a thoughtful address contrasting Liberalism with Socialism.

❧ We want to draw a line below which we will not allow persons to live and labor, yet above which they may compete with all the strength of their manhood. We want to have free competition upwards; we decline to allow free competition downwards. We do not want to pull down the structure of science and civilization but to spread a net on the abyss. ❧

"NO PEACE TILL VICTORY" _1917, London_

The United States entered World War I in April 1917. Shortly thereafter, Prime Minister David Lloyd George asked Churchill to return to the Cabinet as Secretary of State for Munitions and Supplies. To an Anglo-American audience in London in July, Churchill delivered his answer to those who called for a negotiated truce.

❧ Germany must be beaten; Germany must feel like she is beaten. No compromise with the main purpose, no peace till victory, no pact with unrepentant wrong. ❧

"LET US GO FORWARD TOGETHER" _1917, London_

As the end of the war neared, Churchill gave an address in October that would be worthy of his World War II finest. In a peroration that borrowed from his mentor Congressman Bourke Cockran, Abraham Lincoln, and the Bible, Churchill closed his address:

✣ The choice is in our own hands like the Israelites of old, blessing and cursing is not before us. Today we can have the greatest failure or the greatest triumph—as we choose. There is enough for all. The earth is a generous mother. Never, never did science offer such fairy gifts to man. Never did their knowledge and science stand so high. Repair the waste. Rebuild the ruins. Heal the wounds. Crown the victors. Comfort the broken and broken-hearted. There is the battle we have won to fight. There is the victory we have now to win. Let us go forward together. ✣

"DO NOT DELUDE YOURSELVES"   *1932, House of Commons*

In one of his first warnings about Hitler's Germany, Churchill pointed to the revival of a militaristic Germany and the necessity for Britain to rearm.

✣ Now the demand is that Germany should be allowed to rearm. Do not delude yourselves. Do not let His Majesty's Government believe, I am sure they do not believe, that all Germany is asking for is equal status. . . . This is not what Germany is seeking. All these bands of sturdy Teutonic youths, marching through the streets and roads of Germany with the light of desire in their eyes to suffer for the Fatherland are not looking for status. They are looking for weapons and when they have them believe me they will then ask for the return of lost territories or colonies. ✣

THE LOCUST YEARS

From his position as an out-of-government "back bencher," Church-ill condemned the Conservative Party government for failing to rearm.

✛ The government cannot make up their minds, or they cannot get the Prime Minister to make up his mind, so they go in strange paradox, decided only to be undecided, resolved to be irresolute, adamant for drift, solid for fluidity, all powerful to be impotent. So we go on preparing more months and years—precious, perhaps until, to the greatness of Britain—for the locusts to eat. ✛

STAIRWAY DOWN TO DEFEAT

After Hitler's annexation of Austria in August, Churchill, from his "back bench" front corner seat in the House, rose to scold the Conservative government for the five years they had wasted failing to rearm.

✛ For five years I have talked to the House on these matters—not with very great success. I have watched this famous island descending incontinently, fecklessly, the stairway which leads to a dark gulf. It is a fine broad stairway at the beginning, but after a bit the carpet wears. A little further there are only flagstones and a little further on still, these break beneath your feet. . . .

Now is the time at last to rouse the nation. Perhaps it is the last time it can be roused with a chance of preventing war, or with a chance of coming through with victory should our effort to prevent war fail. ⚜

"THE FIRST FORETASTE OF THE BITTER CUP"

1938, House of Commons

In September 1938, Prime Minister Neville Chamberlain returned from Munich after his meeting with Adolph Hitler and announced, "We have achieved peace in our time." Churchill, in an address to the House, denounced those who rejoiced at the settlement.

⚜ They should know that we have passed an awful milestone in our history when the whole equilibrium of Europe has been deranged, and the terrible words have for the time being pronounced against the Western democracies: "Thou are weighed in the balance and found wanting." And do not suppose this is the end.

This is the beginning of the reckoning. This is only the first sip, the first foretaste of the bitter cup, which will be proffered to us year by year, unless by a supreme recovery of moral health and martial vigour, we rise again and take our stand for freedom as in the olden time. ⚜

"LET US TO THE TASK"

1940, Free Trade Hall, London

Just after Britain declared war against Germany, Prime Minister Chamberlain brought Churchill back into the Conservative Cabinet as First Lord of the Admiralty. It was Churchill's second tenure as head of the navy ministry. He had also been First Lord during the early years of World War I.

In January 1940, Churchill would close his speech about the naval war in the North Atlantic with short verbal phrases sublime in their simplicity. It was a foretaste of his later rhetoric as prime minister.

⚜ Come then, let us to the task, to the battle, to the toil—each to our part, each to our station. Fill the armies, rule the air, pour out the munitions, strangle the U-boats, sweep the mines, plough the land, build the ships, guard the streets, succour the wounded, uplift the downcast and honour the brave. Let us go forward together in all parts of the Empire, in all parts of the Island. There is not a week, nor a day, nor an hour to lose. ⚜

"BLOOD, SWEAT AND TEARS"   *1940, House of Commons*

On May 8 the Chamberlain government fell, and King George VI called Churchill to Buckingham Palace with directions to form a new coalition government. Churchill, although not the first choice of the leaders of the Conservative party, was, however, the only Conservative whom the Labor and Liberal party leaders would accept as prime minister.

In his first speech as prime minister, on May 10, Churchill addressed a chamber filled with members, most of whom had rejected Churchill's warnings for the last ten years. The speech became known as "Blood, Sweat and Tears." Actually the original words were "blood, toil, tears and sweat." He had used the phrase in his World War I history—"blood, sweat and tears had spewed upon the plain." Some think he had been inspired by John Donne's poem using those same three words.

✜ I would say to the House, as I said to those who have joined this Government: "I have nothing to offer but blood, toil, tears and sweat."

We have before us an ordeal of the most grievous kind. We have before us many, many long months of struggle and of suffering. You ask what is our policy? I will say: It is to wage war, by sea, land and air, with all our might and with all the strength that God can give us; to wage war against a monstrous tyranny, never surpassed in the dark, lamentable catalogue of human crime. That is our policy. You ask, What is our aim? I answer in one word: Victory—victory at all costs, victory in spite of all terror, victory, however long and hard the road may be; for without victory, there is no survival. Let that be realised; no survival for the British Empire; no survival for all that the British Empire has stood for, no survival for the urge and impulse of the ages, that mankind will move forward towards its goal. But I take up my task with buoyancy and hope. I feel sure that our cause will not be suffered to fail among men. At this time I feel entitled to claim the aid of all, and I say, "Come, then, let us go forward together with our united strength." ✜

On this May Trinity Sunday in the Anglican Church calendar, Churchill delivered a radio address to rally a falling France whose defense at the Sedan had been broken through by the German army. Churchill found reinforcement from the words of the biblical Maccabees, who resisted tyrants who would profane the altar of God's laws.

✥ This is one of the most awe-striking periods in the long history of France and Britain. It is also beyond doubt the most sublime. Side by side, unaided except by their kith and kin in the great Dominions and by the wide Empires which rest beneath their shield—side by side, the British and French peoples have advanced to rescue not only Europe but mankind from the foulest and most soul-destroying tyranny which has ever darkened and stained the pages of history. Behind them—behind us—behind the armies and fleets of Britain and France—gather a group of shattered States and bludgeoned races; the Czechs, the Poles, the Norwegians, the Danes, the Dutch, the Belgians—upon all of whom the long night of barbarism will descend, unbroken even by a star of hope, unless we conquer, as conquer we must; as conquer we shall.

Today is Trinity Sunday. Centuries ago words were written to be a call and a spur to the faithful servants of truth and justice: "Arm yourselves, and be ye men of valour, and be in readiness for the conflict; for it is better for us to persist in battle than to look upon the outrage of our nation and our altar. As the Will of God is in Heaven, even so let it be." ✥

In the aftermath of the miraculous deliverance of the British army in June at Dunkirk, Churchill inspired the House of Commons to thunderous ovation. Members—many with tears in their eyes—stood pounding the benches in front of them when Churchill delivered his defiant words. One member who heard him was the author Harold Nicholson, who wrote to his wife, Vita Sackville-West, that they were "the most magnificent words ever heard in the English language." When President Roosevelt heard the words on radio, he said to his aide Harry Hopkins, "As long as that old bastard is in charge, Britain will never surrender." After hearing this address on the radio, the president of the neutral United States decided to send needed military aid to Britain.

✣ That is the resolve of His Majesty's Government—every man of them. That is the will of Parliament and the nation. The British Empire and the French Republic, linked together in their cause and in their need, will defend to the death their native soil, aiding each other like good comrades to the utmost of their strength. Even though large tracts of Europe and many old and famous States have fallen or may fall into the grip of the Gestapo and all the odious apparatus of Nazi rule, we shall not flag or fail. We shall go on to the end, we shall fight in France, we shall fight on the seas and oceans, we shall fight with growing confidence and growing strength in the air, we shall defend our island, whatever the cost may be, we shall fight on the beaches, we shall fight on the landing grounds, we shall fight in the fields and in the streets, we shall fight in the hills; we shall never

surrender, and even if, which I do not for a moment believe, this
island or a large part of it was subjugated and starving, then our
Empire beyond the seas, armed and guarded by the British Fleet,
would carry on the struggle, until, in God's good time, the new
world, with all its power and might, steps forth to the rescue and the
liberation of the old. ⁂

"WAR OF THE UNKNOWN WARRIORS" *1940, London*

Churchill made this June radio broadcast on the BBC. It was an
address to rally Londoners as they were being subjected to nightly
bombings while they awaited the expected German invasion from
across the Channel.

⁂ Here is the strong City of Refuge, which enshrines the title-
deeds of human progress and is of deep consequence to Christian
civilisation. . . . [W]e await undismayed the impending assault. Perhaps it will come tonight. Perhaps it will come next week. Perhaps it
will never come. We must show ourselves equally capable of meeting
a sudden violent shock or (what is perhaps a harder test) a protracted
vigil. But be the ordeal sharp, or long, or both, we shall seek no terms,
we shall tolerate no parley. We may show mercy—we shall ask for
none. . . .

This is no war of chieftains or of princes, of dynasties or national
ambition. It is a war of people and of causes. There are vast numbers
not only on this island, but in every land who will render faithful
service in this war but whose names will never be known, whose deeds

will never be recorded. This is a war of the unknown warriors; but let us strive without failing in faith or in duty, and the dark curse of Hitler will ever be lifted from our age. ✣

"THEIR FINEST HOUR"

1940, London

When France surrendered in June 1940, Britain was at last alone . . . fighting the war. Churchill would turn the bleakness of challenge into exhilaration. In his closing sentences, he borrowed from the St. Crispin's Day address in Shakespeare's *Henry V*.

✣ What General Weygand called the Battle of France is over. I expect that the battle of Britain is about to begin. Upon this battle depends the survival of Christian civilisation. Upon it depends our own British life, and the long continuity of our institutions and our Empire. The whole fury and might of the enemy must very soon be turned on us. Hitler knows that he will have to break us in this island or lose the war. If we can stand up to him, all Europe may be free and the life of the world may move forward into broad, sunlit uplands. But if we fail, then the whole world, including the United States, including all that we have known and cared for, will sink into the abyss of a new dark age made more sinister, and perhaps more protracted, by the lights of perverted science. Let us therefore brace ourselves to our duties, and so bear ourselves that, if the British Empire and its Commonwealth last for a thousand years, men will still say, "This was their finest hour." ✣

On June 20, 1940, the House of Commons went into secret session to hear the prime minister make a statement following the French debacle. This speech was not recorded, but notes from Churchill's papers exist. The closing phrases seem to echo the crusade ardor of John Bunyan's *Pilgrim's Progress*.

✠ Do not let us lose the conviction that it is only by supreme and superb exertions, unwearying and indomitable, that we shall save our souls alive. No one can predict, no one can even imagine, how this terrible war against German and Nazi aggression will run its course or how far it will spread or how long it will last. Long, dark months of trials and tribulations lie before us. Not only great dangers, but many more misfortunes, many shortcomings, many mistakes, many disappointments will surely be our lot. Death and sorrow will be the companions of our journey; hardship our garment; constancy and valour our only shield. We must be united, we must be undaunted, we must be inflexible. Our qualities and deeds must burn and glow through the gloom of Europe until they become the veritable beacon of its salvation. ✠

"BY SO MANY TO SO FEW"   *1940, House of Commons*

In August 1940, Hitler unleashed the Luftwaffe in a massive bombing attack to bring Britain to its knees. The air invasion was called by Churchill the "Battle of Britain." RAF pilots—mostly in their twenties and younger—manned their Spitfires and Hurricanes to repel the German bombers. In a speech to the House, Churchill paid tribute to the gallant courage of the airmen, employing a phrase he had earlier voiced to the Vice Air Marshall at the time of the German retreat.

⚜ The gratitude of every home in our island, in our Empire, and indeed throughout the world, except in the abode of the guilty, goes out to the British airmen who, undaunted by odds, unwearied in their constant challenge and mortal danger, are turning the tide of the world war by their prowess and by their devotion. Never in the field of human conflict was so much owed by so many to so few. ⚜

"NEVER STOP, NEVER WEARY, AND NEVER GIVE IN"
1940, London

Churchill took to the airwaves in an October broadcast to rally French resistance to the German occupation.

⚜ Never will I believe the soul of France is dead. Never will I believe that her place among the greatest nations of the world has been lost forever. . . .

Remember, we shall never stop, never weary, and never give in, and that our whole people and Empire have vowed themselves to the task of cleansing Europe from the Nazi pestilence and saving the world from the new Dark Ages. . . .

We seek to beat the life and soul out of Hitler and Hitlerism, that alone, that all the time, that to the end. ⚜

"GIVE US THE TOOLS"

This February address was another world broadcast on the BBC. The speech was a direct appeal by Britain to the still neutral United States for military support. In his closing peroration, Churchill refers to a poem which Wendell Willkie, the 1940 Republican candidate for president, had delivered to Churchill from President Roosevelt.

✢ The other day, President Roosevelt gave his opponent in the late Presidential Election, a letter of introduction to me, and in it he wrote out a verse, in his own handwriting, from Longfellow which, he said, "applies to you people as it does to us." Here is the verse:

> . . . Sail on, O Ship of State!
>   Sail on, O Union, strong and great!
> Humanity with all its fears,
> With all the hopes of future years,
>   Is hanging breathless on thy fate!

What is the answer that I shall give, in your name, to this great man, the thrice-chosen head of a nation of 130,000,000? Here is the answer which I will give to President Roosevelt: Put your confidence in us. Give us your faith and your blessings and, under Providence, all will be well. Neither the sudden shock of battle, nor the long-drawn trials of vigilance and exertion will wear us down. Give us the tools and we will finish the job. ✢

"BUT WESTWARD, LOOK, THE LAND IS BRIGHT"

In another world broadcast on the BBC, Churchill reviewed the course of the war that April, noting the bombing of London and the campaign in and around Egypt. He ended with a poem by Arthur Clough that alludes to some victorious progress in the eastern theater of war and then points to the increasing involvement of the U.S.

✤ When we face with a steady eye the difficulties which lie before us, we may derive new confidence from remembering those we have already overcome. Nothing that is happening now is comparable in gravity with the dangers through which we passed last year. Nothing that can happen in the East is comparable with what is happening in the West.

Last time I spoke to you I quoted the lines of Longfellow which President Roosevelt had written out for me in his own hand. I have some other lines which are less well known but which seem apt and appropriate to our fortunes tonight, and I believe they will be so judged wherever the English language is spoken or the flag of freedom flies:

> For while the tired waves, vainly breaking,
>   Seem here no painful inch to gain,
> Far back, through creeks and inlets making,
>   Comes silent, flooding in, the main.
>
> And not by eastern windows only,
>   When daylight comes, comes in the light,
> In front the sun climbs slow, how slowly!
>   But westward, look, the land is bright. ✤

"WE SHALL FIGHT HIM BY LAND"

1941, Chequers, Buckinghamshire, England

While spending a June weekend at the prime minister's weekend residence, Churchill received the news of Hitler's invasion of Russia. He immediately called the BBC to schedule a broadcast from Chequers the following evening. He stayed up most of the night preparing the address.

✣ We have but one aim and one irrevocable purpose. We are resolved to destroy Hitler and every vestige of the Nazi regime. From this nothing will turn us—nothing. We will never parley. We will never negotiate with Hitler or any of his gang. We shall fight him by land. We shall fight him by sea. We shall fight him in the air, until with God's help we had rid the earth of his shadow and liberate its people from his yoke. Any man or state who fights Nazidom will have our aid. Any man or state who marches with Hitler is our foe. ✣

"WHAT KIND OF PEOPLE DO THEY THINK WE ARE?"

1941, Joint Session of Congress, Washington, D.C.

Right after the Japanese attack on Pearl Harbor, Churchill journeyed to Washington to see President Roosevelt. At the same time, he received an invitation to address a joint session of Congress. Churchill was greeted by the U.S. Congress not as a foreigner or as a visiting head of state, but as one of them, a fellow politician with kindred roots and ties. When in the address Churchill asked the rhetorical question about the Japanese "What kind of people do they think we are?" it brought thunderous waves of applause for five minutes.

✠ After the outrages they have committed upon us at Pearl Harbor, in the Pacific Islands, in the Philippines, in Malaya and the Dutch East Indies . . . it becomes difficult to reconcile Japanese action with prudence or sanity.

What kind of people do they think we are? . . .

Duty and prudence alike command . . . that the germ centers of hatred and revenge shall be constantly and vigilantly curbed and bested . . . and that an adequate organization be set up to make sure that the pestilence can be controlled at its earliest beginning before it spreads and rages throughout the entire earth.

It is not given to us to peer into the mysteries of the future. Still I avow my hope and faith sure and inviolate that in the days to come the British and American people will for their own safety and for the good of all walk together in majesty, in justice and in peace. ✠

"SOME CHICKEN, SOME NECK"

December 1941, House of Commons, Ottawa, Canada

While at the White House, Churchill suffered a mild stroke. It was not reported. Despite the seizure, he traveled after Christmas by train from Washington to Ottawa to fulfill an invitation to address the Canadian House of Commons.

The earthy phrase "Some chicken! Some neck!" triggered laughter and an outburst of applause. Churchill, however, was unaware that "neck" at the time was Canadian slang for "brass" or "nerve."

After the speech, Churchill posed for Canadian photographer Youssof Karsh, who plucked the post-oratorical cigar from the prime minister. Disgruntled by the snatching of the cigar, Churchill was caught in a bellicose expression. The result was the most famous photograph of Churchill.

§ We have not journeyed all the way across the centuries, across the oceans, across the mountains, across the prairies because we are made of sugar candy. . . .

We shall never descend to the German and Japanese level but if anybody likes to play rough we can play rough too. Hitler and his Nazi gang have sown the wind; let them reap the whirlwind.

There is no room now for the dilettante, the weakling, for the skirker or the sluggard. The mine, the factory, the dockyard, the salt sea waves, the fields to till, the home, the hospital, the chair of the scientist, the pulpit of the preacher—from the highest to the humblest tasks, are all at equal honor, all have their part to play. . . .

When I warned them [France] that Britain would fight on alone whatever they did, their generals told their Prime Minister and his divided Cabinet, "In three weeks England will have her neck wrung like a chicken." Some chicken! Some neck! §

"THE END OF THE BEGINNING"   *1942, London*

After the victory at El Alamein in November, Churchill said at a Lord Mayor's luncheon:

❖ I have never promised anything but blood, tears, toil and sweat. Now, however, we have a new experience. We have a victory—a remarkable and definite victory.

No, this is not the end. It is not even the beginning of the end. But it is, perhaps, the end of the beginning. ❖

"COMMON CITIZENSHIP"

August 1943, Harvard University, Cambridge, Massachusetts

After meeting with President Roosevelt in Quebec, Churchill took a train southward to Boston to receive an honorary degree at Harvard. In this speech Churchill came closest to proclaiming publicly his private belief in the eventual union between the two countries. The Harvard audience acclaimed the speech, and his wife thought it was his noblest. Perhaps in an implicit concurrence with the Churchill dream, President Roosevelt the next week insisted that the British prime minister preside over a meeting of the combined Chiefs of Staff of both nations in the Cabinet Room of the White House. Roosevelt was vacationing at Hyde Park but invited Churchill to stay in the White House in the president's absence. Even though the countries were wartime allies, it was an unprecedented move.

✢ Twice in my lifetime the long arm of destiny has reached across the ocean and involved the entire life and manhood of the United States in a deadly struggle. . . .

The price of greatness is responsibility. . . . [O]ne cannot rise to be in many ways the leading community in the civilised world without being involved in its problems, without being convulsed by its agonies, and inspired by its causes. . . .

The people of the United States cannot escape world responsibility. . . . We have now reached a point in the journey where . . . it must be world anarchy or world order. . . .

The gift of a common tongue is a priceless inheritance and it may well some day become the foundation of a common citizenship. ✢

"IRON CURTAIN"   1946, *Westminster College, Fulton, Missouri*

In September 1945—in the wake of his devastating defeat—Churchill received an invitation to speak at Westminster College in Fulton, Missouri. His acceptance was, no doubt, spurred by President Truman's addendum to the invitation, saying he would introduce the former prime minister. Churchill carefully prepared for the address. A speech even at a small college in Missouri became a world forum when introduced by the president of the United States. In Washington in late February, Churchill holed up at the British Embassy polishing the draft. Curiously, the most quoted excerpt, which appears here, was added on the train journey; it was not in the advance releases for press offices around the world. The *Washington Post*, for example, missed the famous "iron curtain" lines. Churchill called this March 5 speech "Sinews of Peace," but the world knows it today as the "Iron Curtain" or "Fulton" address. (Churchill took "Sinews" from Cicero's "Sinews of War.")

✤ From Stettin in the Baltic, to Trieste in the Adriatic, an iron curtain has descended across the Continent. Behind that line lie all the capitals of the ancient states of Central and Eastern Europe— Warsaw, Berlin, Prague, Vienna, Budapest, Belgrade, Bucharest and Sofia, all of these famous cities and the populations around them, lie in what I must call the Soviet sphere, and all are subject in one form or another not only to Soviet influence, but to a very high and, in many cases, increasing measure of control from Moscow. ✤

"A SPARK . . . FROM GOD"

1949, Massachusetts Institute of Technology, Cambridge, Massachusetts

Churchill sailed on the *Queen Elizabeth* in 1949 to America to receive a honorary degree at MIT. In March, he landed in New York and spoke there before traveling to Washington for a private meeting with President Truman.

For the thrust of his address he chose the topic of the nature of man and man's ability to rise above the dictates of science.

✤ Laws just or unjust may govern man's actions. Tyrannies may restrain or regulate their words. The machinery of propaganda may pack their minds with falsehood and deny them truth for many generations of time. But the soul of man, thus held in trance or frozen in a long night, can be awakened by a spark coming from God knows where and in a moment the whole structure of lies and oppression is on trial for its life.

People in bondage need never to despair. Let them hope and trust in the genius of mankind. Science no doubt could, if sufficiently perverted, exterminate us all, but it is not in the power of material forces . . . to alter the main element in human nature or restrict the infinite variety of forms in which the soul and genius of the human race can and will express itself. ✥

"A EUROPE STRIVING TO BE REBORN"

1948, The Hague, the Netherlands

In May 1948, Churchill went to the Hague to accept an award from the Dutch government. He used the forum to issue a plea for a united Europe.

✥ A high and solemn responsibility rests upon us here of a Europe striving to be reborn. If we allow ourselves to be rent by pettiness and small disputes, if we fail in clarity of view or courage in action, a priceless occasion may be cast away forever. But if we all pull together and firmly grasp the larger hopes of humanity, then it may be that we shall move into a happier sunlit age, when all the children who are now growing up in this tormented world may find themselves not the victors nor the vanquished in the fleeting triumphs of one country over another, but the heirs of all treasures of the past and the masters of all the science, the abundance and the glories of the future. ✥

"PATIENCE AND ... HOPE"

Churchill had returned to 10 Downing Street after the Conservatives won in the October election. Shortly thereafter, he received an invitation to address a joint session of the U.S. Congress. It would be his second such address. On the last day of 1951, he sailed on the *Queen Mary* to New York.

While in Washington that January, Churchill participated in a plenary session of NATO held at the White House. His address a few days later to Congress focused on the need for America and the West to answer the challenge of Soviet imperialism.

✤ If I may say this, Members of Congress, be careful, above all things. Therefore, not to let go of the atomic weapon until you are sure, and more than sure, that other means of preserving peace are in your hands. . . .

We must not lose patience, and we must not lose hope. It may be that presently a new mood will reign behind the Iron Curtain. If so, it will be easy for them to show it, but the democracies must be on their guard against being deceived by a new dawn.

We seek or covet no one's territory; we plan no forestalling war; we trust and pray that all will come right. . . .

But the great bound forward in progress and prosperity for which mankind is longing cannot come till the shadow of war has passed away. ✤

When George VI died in February 1952, Prime Minister Churchill
delivered a eulogy in a radio broadcast.

❖ The simple dignity of his life, his manly virtues, his sense of
duty—alike as a ruler and a servant—his gay charm and happy
nature, his example as husband and father . . . his courage in peace
and war . . . won the glint of admiration . . . from the innumerable
eyes, whose gaze falls upon the Throne. . . .

He was sustained not only by his natural buoyancy but his Christian faith. During these last months, the King walked with death as if
death were a companion or acquaintance whom he recognized and did
not fear. ❖

"BURDEN OF PEACE"

1953, Conservative Party conference, Margate, England

In June 1953, Prime Minister Churchill suffered a severe stroke. The
illness of Foreign Secretary Anthony Eden and the coronation festivities for Queen Elizabeth II had put severe stress on his constitution.
For a short time, there were doubts about his ability to talk, much less
walk again. There was even fear for his life.

Although the public had little knowledge of his illness, the leaders
of the Conservative Party did, and they thought his retirement
was only a matter of weeks away. Amazingly, Churchill rallied.

At the Conservative Party conference held in October, an appearance if not an address by the party leader was obligatory. He surprised even his family with the vigor of his fifty-minute speech, which mainly addressed foreign policy issues. At the close, he made a personal peroration.

✤ One word personally about myself. If I stay on for the time being bearing the burden at my age it is not because of love of power or office.

I have an ample share of both. If I stay it is because I have a feeling that I may, through things that have happened, have an influence on what I care about above all else, the building of a sure and lasting peace.

Let us go forward together with courage and composure, with resolution and good faith to the end which all desire. ✤

"A HIT-OR-MISS SYSTEM"   *1957, American Bar Association, London*

Churchill chose as the forum for his last great major address the American Bar Association's July meeting in London with its British counterpart. The subject he selected was foreign policy—particularly the United Nations, which he had to come to regard as ineffectual and impotent.

✤ There are many cases where the United Nations have failed. Hungary is in my mind. Justice cannot be a hit-or-miss system. We cannot be content with an arrangement where our system of international laws applies only to those who are willing to keep them. ❧

"A STURDY HORSE" 1959, *Woodford, England*

Churchill's last public speech was to his Woodford constituency, a middle-class suburb of London. A general election in October was called by Prime Minister Harold Macmillan and it was to be a referendum on the economic policies of the Conservative government since 1955.

The Socialists issued a strong challenge, proposing to nationalize industries such as coal mining, shipping, and steel production, which the Conservatives had privatized.

At Woodford Girls School, a crowd gathered to hear their representative deliver the traditional constituency campaign address. At eighty-five, Churchill was a frail man who had to be helped from his chair. The metaphor he chose for private enterprise was striking, and he gestured imitating a rifle and then pumping milk.

❧ Some Socialists see private enterprise as a tiger—a predatory animal to be shot. Others see it as an old cow to be milked. But we Conservatives see it as the sturdy horse that pulls along our economy. ❧

NO "TAME OR MINOR ROLE"

1963, The White House, Washington, D.C.

In April 1963, the U.S. Congress passed a resolution awarding honorary citizenship to Sir Winston Churchill. This came about through the effort of Kay Halle, a friend of the Churchill family. Although he was too feeble to attend, his son Randolph read Churchill's acceptance on the White House steps. In the letter were the defiant words of the old lion.

❧ I reject the view that Britain and the Commonwealth should now be relegated to a tame and minor role in the world. ❧

Coiner of Phrases

Churchill was a master not only in crafting the English sentence but also in the coinage of words. As a writer he was more orator than journalist. He dictated aloud his speeches, articles, and historical chapters, relishing the sound of the words as they rolled off his tongue. He preferred the Old English "feckless" to the Latinate equivalent "ineffectual." He liked "q" or "sq" as pejorative consonants in Anglo-Saxon words such as "squalid" and "queasy." Perhaps that explains how he immediately saw in the name of a Norwegian Nazi collaborator—Quisling—a synonym for traitor.

Some of the words or phrases he minted became the language of world diplomacy, such as "Iron Curtain" and "summit" conference. Others that are not immediately associated with Churchill include "destroyer" and "business as usual."

Shakespeare invented scores of English words by combining prefixes and suffixes with familiar words. Churchill would do the same by coining "unsordid," "unwisdom," and "benignant." By refashioning an old word, he made it leap from the spoken sentence. As a speaker, he knew that words, when spoken, created an effect beyond their literal meeting. Hence, he called the cowardly invasion of France by Mussolini "dastardly."

To his secretary, he would refer to a paper puncher by the onomatopoeic invention "klop." He preferred the hard bite of Anglo-Saxon monosyllabics to Latin polysyllabics ("sweat" instead of "perspiration"). Yet sometimes he resurrected obscure Latinate words for the purpose of comic circumlocution. Two that come to mind are "tergiversation" and "terminological inexactitude."

His memorable word making owes much to Churchill's sense for sound, his massive vocabulary, and his gift for metaphor.

AIRPLANE

✥ Churchill as First Lord of the Admiralty wrote a memo requesting that "aero-plane" be changed to "airplane" and "hydroplane" to "seaplane."

ARBORICIDE

✥ When his wife had chopped down a favorite elm tree at Chartwell, Churchill said to her, "Clemmie, you are guilty of arboricide!"

BENIGNANT

✥ Churchill coined this antonym of "malignant" to describe the growing power of good in the developing Anglo-American relationship.

BLACK DOG

✥ Such was his description of the recurrent moods of melancholy or depression that would seize and overwhelm him.

BLACK VELVET

✢ Such was his frequent phrase for death, a shortened version of his original line "a long sleep on a black velvet pillow."

BLOOD, SWEAT AND TEARS

✢ In his first address as prime minister on May 10, 1940, Churchill said, "I have nothing to offer you but blood, toil, tears and sweat." Perhaps he adapted it from John Donne, who wrote, "Mollify it with thy tears or sweat or blood."

BODYGUARD OF LIES

✢ At the Teheran Conference, Churchill said, "Truth is so precious she must often be attended by a bodyguard of lies." Stalin loved this line.

BOTTLESCAPES

✢ Some of Churchill's first efforts in painting were still-life pictures of tables with a bottle or two and perhaps some fruit. They remained for him a favorite subject. He called them his "bottlescapes." One painting in the 1930s he even titled "Bottlescape."

BOTTLED SUNSHINE

✢ Churchill liked to describe landscapes with this phrase. Churchill's paintings were splashes of vivid color under bright blue skies.

BUSINESS AS USUAL

✢ As First Lord of the Admiralty, Churchill coined this phrase in a World War I speech at the Guildhall in London on November 19, 1914, saying the maxim of the British people is "business as usual."

DESTROYER

✣ As First Lord of the Admiralty in World War I, Churchill designated the word "destroyer" for what had been called "light search and destroy vessel."

END OF THE BEGINNING

✣ "This is not the end, maybe not even the beginning of the end. But it is, perhaps, the end of the beginning." Churchill in a speech in 1942 borrowed the first two lines from Talleyrand before adding his own twist. Talleyrand said of Napoleon's defeat at Waterloo, *"C'est la commencement de la fin."*

FINEST HOUR

✣ Churchill introduced this phrase on June 18, 1940. "[I]f the British Empire and its Commonwealth last for a thousand years, men will still say, 'This was their finest hour.'" It recalls Shakespeare's words in *Henry V*.

IRON CURTAIN

✣ Churchill gave the world the phrase "Iron Curtain" in his speech at Fulton, Missouri, on March 5, 1946. There Churchill told the audience of Westminster College, "From Stettin in the Baltic to Trieste in the Adriatic an iron curtain has descended across the continent." Actually he had used the phrase in 1945 in a letter to President Truman. He also used, before the Truman letter, "iron fence." Although a German reporter also used the phrase "Iron Curtain" in a column in April 1945, there is no evidence that Churchill saw or read the report. At any rate, it was Churchill who popularized the phrase.

NAVAL HOLIDAY

✣ In 1913 Churchill, as First Lord of the Admiralty, proposed to Germany a mutual cessation of shipbuilding. He called for a "naval holiday." The term was picked up by disarmament advocates. In the 1920s, the term became familiar usage in talks on arms limitation between Britain, Japan, France, and America.

PALIMPSEST

✣ Churchill, in 1946, resurrected this rare word. It means a parchment whose previous texts have been insufficiently erased. He used the word to describe inflated, vague jargon of bureaucrats: "One of those rigamaroles and grimaces produced by modern bureaucracy . . . a kind of palimpsest of jargon and affirmative with no breath, no theme and no facts."

PURBLIND

✣ Churchill dug up a Chaucerian adjective meaning "dim-witted" when he described the appeasement-minded intellectuals as "thoughtless dilettantes or purblind worldings."

QUISLING

✣ Vidkun Quisling was the name of a Norwegian Nazi collaborator. In 1943, Churchill made the name a synonym for "traitor": "These vile Quislings within our midst."

SIREN SUIT

✣ Churchill designed a one-piece zip-up suit as quick practical garb to don in case of emergencies, such as when the bombing-alert sirens sounded. He dubbed it his "siren suit."

SPECIAL RELATIONSHIP

⚜ This is the phrase that still describes the Anglo-American political alliance. Churchill coined the phrase in 1946: "Would a special relationship between the United States and the British Commonwealth be inconsistent with our overriding loyalties to the World Organization [U.N.]?"

SUMMIT

⚜ In 1951 Churchill called for "a parley at the summit." The phrase "summit" or "summit conference" caught the world's imagination. It is now a synonym for a diplomatic conference at the highest level.

TERGIVERSATION

⚜ Churchill reintroduced this word for changing one's position or stand—or, more precisely, for reversing one's original position and then switching back again.

TERMINOLOGICAL INEXACTITUDE

⚜ Churchill in 1906 coined this euphemism for "lie": "Perhaps we have been guilty of some terminological inexactitudes." He was referring to government denials of exploitation of Chinese coolie labor in South Africa.

TRIPHIBIAN

⚜ Churchill coined this phrase to describe a warrior proficient on land, air, and sea: "Lord Mountbatten is a triphibian."

UNSORDID

✢ Churchill combined a negative prefix with a familiar word to describe the Lend Lease program: "the most unsordid act." Nowhere else is this word found in literature or speeches. By negativizing "sordid," he coined an arresting word to suggest magnanimity.

UNWISDOM

✢ Again Churchill fashioned an arresting word with a negative prefix. In *The Second World War* he writes about policies before World War II as "unwisdom prevailed."

YEARS OF THE LOCUST

✢ Churchill borrowed from a verse in Job ("That which the locust has eaten") to describe the 1930s, when a somnolent and apathetic Britain failed to rebuild its defense against Nazi Germany. He called the 1930s "the years the locusts have eaten."

Saints and Sinners

In his more than half a century on the world stage, Churchill had the opportunity to meet, observe, and get to know more leaders than any man in history. With the exception of Coolidge and Harding, he met every U.S. president from McKinley to Kennedy. He also knew and corresponded with Richard Nixon.

As the Cabinet minister in two world wars, he met and knew Woodrow Wilson and Franklin Roosevelt, Georges Clemenceau and Charles de Gaulle.

To his profession of politics he brought his skills as a journalist. He could pierce through the cant and hypocrisy of politicians and separate the great from those who merely postured.

First as a soldier, then as a reporter, and finally as a politician, he found himself fascinated by history. Soon he turned his talents toward biography, and then history.

If historians are divided simplistically into two camps—those who believe "man makes history" and those who believe "history makes the man"—he was the most fervent disciple of the former school. With his confidence in his own destiny, he had to believe that leaders, for good or evil, could shape the course of history.

Churchill's knowledge of history enabled him to understand more perceptively his own colleagues, rivals, and foes in the political and world arena.

HAROLD ALEXANDER
(*Earl Alexander of Tunis, British field marshall and general*)

❧ He is no glory-hopper.

❧ Cool, gay, comprehending all, he inspired quiet, deep confidence in every quarter.

QUEEN ANNE
(*Queen of England, 1702–1714*)

❧ She moved on broad homely lines. . . . She was not very wise nor clever but she was very like England.

HERBERT ASQUITH
(*Liberal party leader and prime minister, 1912–1917*)

❧ Asquith's opinions in the prime of his life were cut in bronze.

CLEMENT ATTLEE
(*Socialist party leader and prime minister, 1945–1951*)

❧ He is a modest man with much to be modest about.

❧ He is a sheep in sheep's clothing.

STANLEY BALDWIN
(*Conservative Party leader, 1924–1930 and prime minister, 1923–1924; 1924–1929; 1935–1937*)

❧ Occasionally he stumbled over the truth but hastily picked himself up as if nothing had happened.

❧ He has his ear so close to the ground that he has locusts in it.

ARTHUR JAMES BALFOUR
(*Conservative Party leader, prime minister, 1902–1905 and foreign secretary 1916–1922*)

✣ If you wanted nothing done, Arthur Balfour was the best man for the task. There was no equal to him.

LORD BEAVERBROOK
(*Sir Max Aitken, British press magnate and member of wartime Cabinet*)

✣ He is a foul-weather friend!

✣ That actor Edward G. Robinson looks just like Max.

ANEURIN BEVAN
(*British leftist and socialist*)

✣ He is a merchant of discourtesy.

RUPERT BROOKE
(*British poet who died in World War I*)

✣ A voice had become audible, a note had been struck, more true, more thrilling, more able to do justice to the mobility of youth in arms engaged in this present war ... the voice has been swiftly engaged in this present war and memories remain but they will linger. . . .

R. A. ("RAB") BUTLER
(*Conservative Chancellor of the Exchequer, 1951–1955*)

✣ I am amused by the Chancellor of the Exchequer. He is always patting himself on the back, a kind of exercise that contributes to his excellent physical condition.

JOSEPH CHAMBERLAIN
(British politician and reformer, Radical parliamentary leader)

⚜ Joe Chamberlain loves the working man. He loves to see him work.

NEVILLE CHAMBERLAIN
(Conservative Party leader, prime minister, 1938–1940)

⚜ He has a lust for peace.

⚜ He was given the choice between war and dishonor. He chose dishonor and he will have war anyway.[1]

CHARLIE CHAPLIN
(American movie actor)

⚜ He is a marvelous comedian—bolshy[2] in politics and delightful in conversation.

LORD RANDOLPH CHURCHILL
(Winston's father, Conservative party leader, and Chancellor of the Exchequer, 1888–1894)

⚜ Most people grow tired before they are over-tired. But Lord Randolph was of the temper that gallops until it falls.

GEORGES CLEMENCEAU
(French prime minister, 1914–1921)

⚜ A life of storm, from beginning to end; fighting all the way; never a pause, never a truce, never a rest.

---

[1] Just after the Munich settlement with Hitler in 1938.
[2] Bolshevik.

SIR STAFFORD CRIPPS
(Socialist Chancellor of the Exchequer, 1945–1951, a teetotaler and vegetarian)

❧ There but for the grace of God goes God.

❧ His chest is a cage in which two squirrels are at war—his conscience and career.

CHARLES DE GAULLE
(leader of the free French and president of France, 1959–1970)

❧ He looks like a female llama who has just been surprised in her bath.

❧ We all have our crosses to bear; mine is the cross of Lorraine.

❧ In the last four years I have had many differences with General de Gaulle, but I have never forgotten, and can never forget, that he stood forth as the first eminent Frenchman to face the common foe in what seemed to be the hour of ruin of his country, and possibly of ours.

❧ France without an Army is not France—de Gaulle is the spirit of that Army. Perhaps the last survivor of a warrior race.

❧ General de Gaulle shone as a star in the pitch-black night.

❧ *L'homme du destin* ("Man of destiny").

JOHN FOSTER DULLES
(U.S. Secretary of State, 1953–1959)

❧ He is the only bull I know who carries his own china closet with him.

❧ Dull, duller, Dulles.

ANTHONY EDEN
(Conservative Party leader, 1940–1945, foreign secretary, 1951–1955 and prime minister, 1955–1957)

✢ He is the one fresh figure of first magnitude arising out of a generation which was ravaged by the war [World War I].

DWIGHT DAVID ("IKE") EISENHOWER
(general and U.S. president, 1953–1961)

✢ He is a prairie prince.

✢ A man who set the unity of the Allied Armed Forces above all nationalistic thought.

✢ A broad-minded man, practical, serviceable, dealing with events as they came, in cool selflessness.

✢ Never have I seen a man so staunch in pursuing the purpose at hand, so ready to accept responsibility for misfortune or so generous in victory.

ELIZABETH II
(Queen of England, 1952–   )

✢ Lovely, inspiring. All of the film people in the world if they had scoured the globe could not have found anyone so suited to the part.

MAHATMA GANDHI
(Indian leader)

✢ A seditious Middle Temple lawyer, now posing as a fakir of a type well known in the East, striding half-naked up the steps of the Vice Regal Palace to parley with the representative of the King

CHARLES ("CHINESE") GORDON
(*general, martyr of Khartoum massacre*)

✢ A man careless alike of the frown of men or the smiles of women, of life or comfort, wealth or fame.

EARL MARSHAL DOUGLAS HAIG
(*World War I British General*)

✢ [Like] a surgeon . . . sure of himself, stead of poise, knife in hand, intent upon the operation; entirely removed from the agony of the patient. . . . He would operate without excitement . . . and if the patient died, he would not reproach himself.

EARL OF HALIFAX
(*British foreign secretary, 1939–1940*)

✢ Halifax's virtues have done more harm than the vices of hundreds of other people.

PAUL VON HINDENBERG
(*president of Germany, 1928–1933*)

✢ Hindenberg had nothing to learn from modern science and civilization except its weapons; no rule of life but duty. . . .

In the last phase we see the aged President having betrayed all the Germans who had re-elected him to power, joining reluctant and indeed contemptuous hands with the Nazi leader. There is a defense for all this . . . he had become senile.

ADOLF HITLER
(*dictator of Nazi Germany, 1933–1945*)

✤ Into that void strode a maniac of ferocious genius of the most virulent hatred that has ever corroded the human breast . . . Corporal Hitler.

✤ This wicked man, the repository and embodiment of many forms of soul-destroying hatred, the monstrous product of former wrongs and shames.

✤ This bloodthirsty guttersnipe.

✤ Hitler and his Nazi gang have sown the wind; let them reap the whirlwind.

HARRY HOPKINS
(*President Franklin Roosevelt's principal aide*)

✤ I dub you "Lord Root of the Matter."

✤ Extraordinary. . . . In the history of the United States few brighter flames have burned.

LORD JELLICOE
(*British admiral in World War I, Commander of the British Home Fleet*)

✤ He was the only man on either side who could lose the war in an afternoon.

JOAN OF ARC

✤ There now appeared on the ravaged scene an Angel of Deliverance, the noblest patriot of France, the most splendid of her heroes, the most beloved of her saints, the most inspiring of all her memories, the peasant Maid, the ever-shining, ever glorious Joan of Arc.

KING FAROUK OF EGYPT (1944–1953)

⚜ King Farouk was wallowing like a sow in a trough of luxury.

KING JOHN
(*King of England, 1199–1216*)

⚜ When the long tally is added, it will be seen that the British nation and the English-speaking world owe far more to the vices of John than to the labors of virtuous sovereigns.

LORD KITCHENER
(*World War I British general*)

⚜ He may be a general but never a gentleman.

T. E. LAWRENCE
(*"Lawrence of Arabia"*)

⚜ He was not in complete harmony with the normal.

⚜ He had the art of backing reluctantly into the limelight. He was a very remarkable character and very careful of that fact.

VLADIMIR LENIN
(*Communist leader of the Soviet Union*)

⚜ Implacable vengeance rising from a frozen pity. His sympathies cold and wide as the Arctic Ocean; his hatreds tight as the hangman's noose. His purpose to save the world; his method to blow it up.

⚜ It is with a sense of awe that they [the Germans] turned upon Russia the most grisly of all weapons. They transported Lenin in a sealed truck like a plague bacteria from Switzerland into Russia.

DAVID LLOYD GEORGE
(British prime minister, 1916–22, in World War I)

‡ He could talk a bird out of a tree.

‡ He is not against the social order but only those elements of social order that get in his way.

‡ His warm heart was stirred by the many perils which beset the cottage homes, the health of the breadwinner, the fate of his widow, the nourishment and upbringing of his children. The meager and haphazard provision of medical treatment. . . . All this excited his wrath. Pity and compassion lent their powerful wings.

LOUIS XIV
(King of France, 1643–1715)

‡ Petty and mediocre in all except his lusts and power, the Sun King disturbed and harried mankind during more than fifty years of arrogant pomp.

LORD MACAULAY
(British historian)

‡ It is beyond our hopes to overtake Lord Macaulay . . . we can only hope that truth will follow swiftly enough to fasten the label "Liar" to his genteel coattails.

RAMSAY MACDONALD
(Socialist leader, 1922–1924 and prime minister, 1930–1932)

‡ He has the gift of compressing the largest amount of words into the smallest amount of thought.

‡ The greatest living master of falling without hurting himself.

VYACHESLAV MOLOTOV
(*Soviet Union foreign secretary*)

✤ I have never seen a human being who more perfectly represented the modern conception of a robot.

VISCOUNT BERNARD ("MONTY") MONTGOMERY
(*field marshal and general*)

✤ In defeat, indomitable. In advance, invincible. In victory, insufferable.

LORD MORLEY
(*Liberal Party leader and historian*)

✤ The old world of culture and quality . . . was doomed; but it did not lack its standard-bearer.

LORD LOUIS MOUNTBATTEN
(*Earl Mountbatten of Burma, admiral and viceroy of India*)

✤ He is a triphibian—equally at home on land, sea or air and he has experienced a bit of fire too.

BENITO MUSSOLINI
(*leader of Fascist Italy*)

✤ The hyena in its nature broke all bounds of decency and common sense.

✤ This tattered lackey at [Hitler's] tail . . . this whipped jackal frisking at the side of the German tiger.

NICHOLAS II
(*Czar of Russia*)

✤ He was neither a great captain nor a great prince. He was only a true simple man of average ability, of merciful disposition, upheld in all his daily life by his faith in God.

LORD OLIVIER (SIR LAURENCE)
(British actor)

✤ He has a myriad of the Bard's words in his brain.

RICHARD THE LION-HEARTED
(King of England)

✤ His life was one magnificent parade, which when ended left only an empty plain.

SIR WILLIAM PITT (EARL OF CHATHAM)
(Eighteenth-century British prime minister)

✤ His policy was a projection on a vast screen of his own aggressive personality. . . . To call into life and action the depressed and languid spirit of England, to weld her resources of wealth and manhood into a single instrument of war . . . to conquer and command and never to count the cost whether in blood or gold——this spirit he infused into every rank of his countrymen.[3]

ERWIN ROMMEL
(German field marshal and general)

✤ We have a very daring and skillful opponent against us, and, may I say across the havoc of war——a great general.[4]

---

[3] Churchill wrote these words in 1939—before Hitler attacked Poland. The description is prophetic about the role he was himself soon to play.
[4] These words, uttered in a House of Commons address in 1942, were much criticized.

FRANKLIN D. ROOSEVELT
(*U.S. president, 1932–1945*)

 ⚜ Meeting him was like opening your first bottle of champagne.

 ⚜ That great man whom destiny has marked for the climax of human fortune.

 ⚜ He was the greatest friend Britain ever had.

 ⚜ He died on the wings of victory, but he saw them and heard them beating.

ELEANOR ROOSEVELT
(*First Lady, 1932–1945*)

 ⚜ You have certainly left golden footprints behind you.

LORD ROSEBERY
(*prime minister, 1896–1899 and Liberal Party leader*)

 ⚜ He flourished in an age of great men and small events.

GEORGE BERNARD SHAW
(*British playwright*)

 ⚜ Few people practice what they preach and none less so than George Bernard Shaw. . . . Saint, sage and clown; venerable, profound and irresistible.

HARRY S. TRUMAN
(*U.S. president, 1945–1952*)

 ⚜ He seems a man of exceptional character and ability . . . [possessing] simple and direct methods of speech and a great deal of self-confidence and resolution.

✣ He is a man of immense determination. He takes no notice of delicate ground, he just plants his foot down firmly on it.

✣ We not only speak the same language, we think the same thoughts.

GEORGE WASHINGTON
✣ The military fame of George Washington would rest not on the Chesapeake but on the Delaware. It was that marvelous bitter nerve-wracking campaign that revealed the fortitude and constancy of the American leader.

WOODROW WILSON
(*U.S. president*)
✣ Peace and goodwill among all nations abroad, but no truck with Republicans at home.

✣ If Wilson had been either simply an idealist or a caucus politician, he might have succeeded. His attempt to run the two in double harness was the cause of his undoing.

DUKE OF WINDSOR (KING EDWARD VIII, 1936)
✣ A little man dressed up to the nines.

Escapades and Encounters

In a sense, it is misleading to recount many of the delicious anecdotes about Winston Churchill. They could serve to distract from the greatness of his accomplishments, for character can slip into caricature. The result may be a distorted emphasis on the bibulous Churchill, which is as wrong as seeing Benjamin Franklin only as an amiable lecher. Churchill was more than a lovable character.

However, the knack of taking one's cause but not oneself seriously is surely a trait of leadership sorely missed in the presidents of today. The humorless politician may smack of the pompous prig or ideological fanatic. Such personalities are, in the long run, ill-suited to the compromise and tolerant give-and-take of democratic life. The lusty wit of Churchill was an antidote to cant and political posturing.

The Churchill stories—even if not all of them can be documented—reveal the colossal richness of his personality. Perhaps some of the stories that follow are apocryphal or have been altered slightly to heighten a point. Many of them were told me by friends and associates of Churchill and British members of Parliament, and cannot be authoritatively supported. But the humorous accounts of Churchill's encounters with the great and not so great are more than just insights into his character. Many of them touch on the human condition in general, such as the vices of hypocrisy and arrogance or the vicissitudes of marriage and old age.

ALCOHOL AND ALLAH

The last great cavalry charge in British military history in 1897 happened to be Churchill's first, and Churchill's alertness as a scout in reporting the impending raid of "the Whirling Dervishes" to General Kitchener might have spelled the difference between victory and defeat.

Though many of his fellow Lancers fell, Churchill survived the onslaught of the scimitar-wielding Islam fanatics. It was cause for celebration but, alas, no fuel for toast could be found in the empty mess hall cellar. Undaunted, Churchill mounted his steed for a jaunt to the Nile River. There he hailed a patrolling British gunboat and cried out for liquid relief. A bottle of bubbly was tossed out by a sympathetic officer[1], and Churchill got off his horse and waded in to retrieve the prize.

As he raised his glass with fellow officers, he intoned, "It is altogether fitting to imbibe what the dictates of our foes proscribe."

---

[1] The young naval officer was David Beatty. Perhaps it is only coincidental that when Churchill became First Lord of the Admiralty in 1912, he promoted Beatty over the ranks of many more senior to be Admiral.

ANOTHER MAN'S POISON

❧ Nancy Astor was a native Virginian who became Britain's first woman member of the House of Commons. In the 1930s she headed a clique in the House of Commons that found something to admire in Hitler's Germany. Churchill described an Astorite as an appeaser "who feeds the crocodile hoping that it will eat him last." One time shortly thereafter, Churchill found himself at Cliveden, the Astor mansion.

After dinner Lady Astor presided over the pouring of coffee. When Churchill came by, she glared and said, "Winston, if I were your wife, I'd put poison in your coffee."

"Nancy," Churchill replied to the acid-tongued woman, "if I were your husband, I'd drink it."

AT YOUR PLEASURE

❧ At the end of one speech in the 1930s, a younger member of Parliament who had heard Churchill's address asked him, "Mr. Churchill, why do you never begin your address saying 'It is a pleasure to . . .'?"

Churchill answered, "It may be an honor, but never a pleasure. There are only few things from which I derive intense pleasure, and speaking is not one of them."

BARE FACTS

❧ During Churchill's first visit to the White House as prime minister in December 1941, he was surprised one morning with an unexpected visit by President Roosevelt to Churchill's guest chamber, the Monroe Room. FDR, who was propelling himself in his wheelchair, opened the door to find his guest stark naked and gleaming pink from his morning bath. Faced with this vision, the American president put his chair into reverse. But Churchill stopped him, saying, "Pray enter. His Majesty's First Minister has nothing to hide from the president of the United States."

BIRDS OF A FEATHER

❖ In 1944, Churchill made a Christmas visit to Athens. In Athens, the Prime Minister met General Alexander. He asked Alexander what kind of a man was Damaskenos, the Orthodox Archbishop of Greece.

"Is he one of those priestly ascetics concerned only with spiritual grace or one of those crafty prelates concerned rather with temporal gain?"

"Regrettably," replied Macmillan, "the archbishop seems to be one of the latter."

"Good," replied Churchill, rubbing his hands. "Then he is just our man."

BLEAK HOUSE

❖ As leader of the Opposition after World War II, Churchill soon found that the only thing as obnoxious as Socialists' planning was their jargon. The poor became "the lower income disadvantaged" or "marginal stipend maintainers."

Churchill bitterly reacted to their description of "house" or "home" as "local accommodation unit." He said in the House of Commons: "Now we will have to change that old favorite song 'Home, Sweet, Home' to say 'Local Accommodation Unit, Sweet

Local Accommodation Unit—there's no place like Local Accommodation Unit.' "

BLOWHARD

✛ One of the admirals who was violently opposed to Churchill's reforms of the Royal Navy before World War I was a windbag named Lord Charles Beresford.

Churchill reacted to one of his choleric outbursts by saying, "He can be described as one of those orators who, before he gets up, does not know what he is going to say; when he is speaking, does not know what he is saying; and when he has sat down, does not know what he has said."

BOOK VALUE

✛ Lord Londonderry, though a cousin of Churchill's, was a frequent political foe. A pacifist, Londonderry wrote articles and pamphlets to promote his views.

One day, armed with his new works, he encountered Churchill in the street.

"Winston, have you read my latest book?"

"No," replied Churchill. "I only read for pleasure or for profit."

BORE AND SNORE

✛ During a long session in the House of Commons, one of Churchill's Socialist opponents was droning on in a tedious discourse. Churchill reacted by slumping in his seat and closing his eyes. Noting the nap by Churchill, the speaker said, "Must the right honorable gentleman fall asleep when I am speaking?"

Churchill blithely replied, "No, it is purely voluntary."

BRUSH-OFF

☙ After his escape from capture in the Boer War, the twenty-six-year-old Churchill won a seat in Parliament. To make himself look older, he grew a mustache. A woman acquaintance who was not enthusiastic about his independent political views encountered him at a dinner party.

"Winston," she scolded, "I approve of neither your politics nor your mustache."

"Madam," replied Winston, "you are not likely to come in contact with either."

BUTT-ERING UP

☙ Charles de Gaulle, the self-appointed head of the Free French, seethed at Allied discussions with the Vichy France rulers of North Africa. Yet any bloodless invasion of Morocco and Algeria depended on the compliance of both Vichy leaders and de Gaulle. When a British diplomat urged the prime minister to meet with de Gaulle and try to soothe his Gallic pride with flattery, Churchill said, "I'll kiss him on both cheeks—or if you prefer, on all four."

CHANCES ARE ...

☙ On November 30, 1949, Churchill celebrated his seventy-fifth birthday. A photographer at the birthday event said, "I hope to have the opportunity to take your photograph on your hundredth birthday."

Churchill looked the photographer up and down.

"I don't see why not, young man," he replied. "You look healthy enough."

CHEAP AND NASTY

✤ When the first destroyers arrived in the fall of 1940 under America's Lend-Lease Program to Great Britain, Prime Minister Churchill went to inspect them. He was joined by FDR's right-hand man, Harry Hopkins.

Churchill looked at the decidedly overaged rustbuckets and grumbled in a whisper, "Cheap and nasty."

Hopkins, who was startled by the remark, queried, "What was that?"

Churchill amended aloud, "Cheap for us and nasty for the Germans."

CHINESE TORTURE

✤ In 1942, when Churchill's leadership in the war effort was coming under attack, Churchill had this proposal for his critics:

"There was a custom in ancient China that anyone who wished to criticize the government had the right to memorialize the emperor, provided he followed up by committing suicide. Very great respect was paid to his words and no ulterior motive was assigned. That seems to me to have been from many points of view a very wise custom, but I certainly would be the last to suggest that it should be made retroactive."

CLEAN OF TARTS?

✤ Churchill's first public speech was in 1894. It was not, however, a "maiden" address. In fact, he delivered an exhortation in support of unvirginal pursuits. The Women's Entertainment Protection League had mounted a campaign to remove the solicitation by "ladies of the evening" at the music halls. At the insistence of this bluenose lobby, the Empire Theatre erected canvas partitions to prevent streetwalkers from catching the eyes of young blades drinking at the theater's bar.

Cadets at Sandhurst, as well as other young officers, during breaks between the variety skits, were accustomed to mixing their booze with some banter with the wenches who strolled by. They enlisted subaltern Churchill's help to resist the moral crackdown.

Churchill, along with some like-minded officers, bought tickets to a performance at the Empire. Then, during an intermission, he sounded the signal for action: "Charge the barricades," he roared. The canvas was ripped down and Churchill mounted the stage and inveighed against "the prowling prudes" who would deny Londoners of their civil liberties. His final presentation closed with the words, "Gentlemen, you have seen us pull down the barricades. See that you pull down those who are responsible for them at the coming election."

COLD AND HOT

❦ During World War II, an aide read to Churchill an account in a tabloid: A seventy-five-year-old man on a cold January day—with the temperature thirty-five degrees below freezing—had propositioned a nineteen-year-old girl to have sex on the grass in Hyde Park.

Churchill replied to the aide, "Over seventy-five! Below-zero temperature! It makes you proud to be an Englishman!"

COLD SOUP

❦ As a young man, Churchill was asked by a friend about the London dinner party he had attended the night before.

"Well," replied Churchill, "it would have been splendid . . . if the wine had been as cold as the soup, the beef as rare as the service, the brandy as old as the fish, and the maid as willing as the Duchess."

CONTAINMENT POLICY

⚜ A Labourite, in the midst of his attack on the policies of the Conservative government, suddenly lost his train of thought. Churchill interceded with this sympathetic suggestion:

"My right honourable friend should not develop more indignation than he can contain."

COUNT YOUR BLESSINGS

⚜ During a boring discourse when a Labourite academic was droning on about Socialist theory, Churchill noted an elderly member of Parliament straining with his ear trumpet to listen.

"Who is that fool who is denying his natural advantages?" he demanded.

COVER ONE'S ASS

⚜ As leader of the Conservative Party Opposition in the 1950s, Churchill grew quite restless and bored as a Socialist academic pontificated against the big business and the rich.

Churchill replied, "Verbosity may be the long suit of the honourable gentleman, but it's not long enough to cover his ass-ininity."

CREEPING SENILITY

⚜ When a veteran parliamentarian rambled disjointedly against Churchill's wartime policies, the prime minister delivered this medical assessment:

"I must warn him that he will run a very grave risk of falling into senility before he is overtaken by age."

A "DANIELS" COME TO JUDGMENT

✣ On March 2, 1946, President Truman greeted Churchill at Union Station, where the presidential party and the British guests boarded the private Presidential train for the trip for Fulton, Missouri. Churchill was to speak the next day at Westminster College, on which occasion he would deliver his famous "Iron Curtain" address. As soon as the train steamed its way from the station, Truman said, "Churchill, let's have some whiskey!"

"Capital idea," replied Churchill.

The steward brought in a tray of glasses with ice and a bottle of Jack Daniels to the lounge car.

A dismayed Churchill looked at the bottle and exclaimed, "This isn't whiskey—it's bourbon!"

And Churchill then pulled the emergency cord. The train came to a screeching halt just outside Washington at Silver Spring, Maryland, where a case of Johnnie Walker Red Label was brought to the rescue.

DEAD BIRDS

✣ While sitting on a platform waiting to speak, the seventy-eight-year-old Churchill was handed a note by an aide. Churchill glanced at the message, which advised, "Prime Minister—your fly is unbuttoned."

Churchill then scrawled on the bottom of the note and passed it back. It read, "Never fear. Dead birds do not drop out of nests."

DEAD CERTAINTY

✢ Although Stanley Baldwin was popular with the people, Churchill believed that the Conservative leader's views would eventually drag down the Conservative Party.

Once when Baldwin was ill, a Conservative Member lamented the fact to Churchill.

"What would happen if our beloved Stanley would die?"

Churchill replied, "Embalm, cremate and bury. Take no chances."

DEAF AND DUMB

✢ In the 1960s the feeble Churchill only occasionally attended the House of Commons as the member representing the London bedroom constituency of Woodford.

As Churchill lolled on the bench, he overheard some members talking behind of him: "Poor fellow—he's a bit ga-ga, you know—he can't understand a thing that's going on."

Churchill turned around and, fixing them with a stare, roared, "Yes, and they say the old boy's quite deaf, too."

DEEP COVER

✢ Churchill was shattered when his wartime victory over the Germans was immediately followed by the overwhelming defeat of his government.

His wife tried to console him: "Dear, it may have been a blessing in disguise."

Churchill grumbled in response, "If that is so, it is certainly effectively disguised."

DELAY OF GAME

✧ When Churchill visited New York in the early 1930s, he was taken to his first American football game at Columbia University. Asked for his comment, he replied:

"Actually, it is somewhat like rugby. But why do you have to have all those committee meetings?"

THE DEVIL HIS DUE

✧ When Hitler invaded Russia in June 1941, Churchill immediately urged support for the Russian army. His aide John Colville asked how he reconciled such a request with his damnation of the Soviet Union dictatorship over the years.

He answered, "If Hitler invaded the realms of Hell, I would at least make a favorable reference to the devil in the House of Commons."

DEVIL'S ADVOCATE

✧ When Churchill first was a candidate in 1900, he did some door-to-door canvassing and things were going pretty well, he thought, till he came to the house of a grouchy-looking fellow. After Churchill's introduction of himself, the fellow said, "Vote for you? Why, I'd rather vote for the devil!"

"I understand," replied Churchill. "But in case your friend is not running, may I count on your support?"

DIVINE DESTINY

✧ In the early days of the Germans' blitzkrieg on London, Churchill motored hurriedly to Canterbury to see that proper precautions were taken for the protection of the famous cathedral there.

He reassured the Archbishop of Canterbury, "We have bolstered the edifice and approaches with sandbags to spare. Every device known to man has been applied. No matter how many close hits the Nazis may make, I feel sure the cathedral will survive."

"Ah, yes, close hits," said the archbishop gloomily, "but what if they score a direct hit upon us?"

"In that event," decided Churchill with some asperity, "you will have to regard it, my dear archbishop, as a divine summons."

DRINK ONLY TO THINE EYES

✢ In the 1920s Britain, like America, enacted women's suffrage. The first woman to be a Member of Parliament was the Virginia-born Lady Astor. She entered the House of Commons as a Conservative member about the same time Churchill switched back to that party from being a Liberal.

Astor, a strict Christian Scientist, was a teetotaler, and not the least factor in her distaste for Churchill was his drinking habits.

Once while Churchill was addressing the House, he spotted Nancy entering the chamber. Churchill then paused to fill his water glass, saying, "It must be a great pleasure for the noble lady, the member of the Sutton Division of Plymouth, to see me drink water."

DRIP

✢ Once a Socialist member of Parliament interrupted a Churchill speech with a rebutting fact.

Churchill shrugged it off, saying, "I do not challenge the honourable gentleman when the truth leaks out of him by accident from time to time."

DOGS OF WAR

✤ During World War II, Churchill's poodle, Rufus, wandered into the midst of a Cabinet Room meeting at 10 Downing Street.

At the empty chair next to his master, who was presiding, Rufus stood in expectant attention.

"No, Rufus," he counseled his canine associate. "I haven't yet found it necessary to ask you to join the wartime Cabinet."

DO OR DIE

✤ In World War I, Churchill was appointed by David Lloyd George to be Secretary of State for Munitions. At the same time, he was the world's first "air minister" as Secretary of State for Air.

Rather than wait for telegrams from the Western Front in France, Churchill would pilot himself over during the morning to inspect where supplies were needed and then fly back that afternoon to report to the House of Commons. His wife, however, took a dim view of his daily air commuting.

When his biplane crashed one afternoon, Churchill managed to emerge unscathed. But he would never fly again. The threat from his wife was:

"Either Clementine or the aeroplane!"

DOWN IN FLAMES

✤ A younger member of Parliament once sought the great orator's advice on speaking.

"Mr. Churchill," he began, "you heard my talk yesterday. Can you tell me how I could have put more fire in my speech?"

"What you should have done," replied Churchill "is to have put your speech into the fire."

DROWNED RAT
✢ In 1945, when a Labourite member crossed the aisle to become a Liberal, Churchill had this comment: "That's the first I've heard of a rat swimming towards a sinking ship."

DRY RUN
✢ One afternoon, in the House of Commons, a Labourite Cabinet Member was delivering a dreary discourse on Socialist accomplishments. Churchill interrupted the tedious talk. With mock sympathy, he consoled the speaker.

"I can well understand the honourable member's wishing to speak for practice. He needs it badly."

EARTH BOUND
✢ During the Festival of Britain in 1951, Prime Minister Churchill visited the Dome of Discovery. He was taken up in a lift to a telescope. "Mr. Churchill," he was told, "here you can view outer space."

"No thank you," replied Churchill, "I'm more interested in what is happening on Earth."

EQUAL TIME
✢ In May 1955, a debate was conducted over the BBC entitled "Christianity vs. Atheism." Churchill objected to the programming.

A BBC spokesman responded, "It is our duty to truth to allow both sides to debate."

Churchill shot back, "I suppose then that if there had been the same devices at the time of Christ, the BBC would have given equal time to Judas and Jesus."

EXHILARATING CIRCUMSTANCES

✢ Though a fledgling officer in India, Churchill yearned for a political career like his father had. But members of the House of Commons in those days received no salaries and Churchill had no independent income. His solution, he told his mother, was to seek adventure, find glory, and write about it.

While stationed in Bangalore, India, he volunteered during vacation leave for a "search and seizure" mission five hundred miles north on the Afghan border. Pathan tribesmen from their ridge position wiped out his detachment. Churchill soon found himself all alone, pinned down by their continuing fire. With his pistol he stood off his pursuers and eventually escaped. When he returned to Bangalore, he wrote down his notes of his first military adventure. "I found that there is nothing more exhilarating than to be shot at without result."

FACADE OF FINANCE

✢ At a diplomatic reception in London before World War I, an Italian military attaché noticed a medal worn by the Luxembourg minister and asked about it. The Luxembourg diplomat replied stiffly that it was an ancient order called the Royal Admiralty Cross.

The Italian diplomat then remarked to Churchill, who was then First Lord of the Admiralty in Britain, "Mr. Churchill, can you believe that Luxembourg has an admiralty award and they don't even have a navy?"

Churchill replied, "Why shouldn't they have an admiralty? You in Italy, after all, have a Ministry of Finance—yet you don't have a treasury!"

FAIR MEMORY

✢ One time in the 1920s, Churchill was accosted by one of the audience after a talk. The man grabbed Churchill's sleeve and said, "Mr. Churchill. Perhaps you don't remember me, but I was the innkeeper at the King's Arms when you were the member from Dundee before the Great War. But you might recall Katie, a buxom lass of fair face and full figure—who was the barmaid at the King's Arms."

"No doubt I would have remembered if—in the King's Arms—I had been in Katie's arms."

FALLEN FLAT

✢ While vacationing in the south of France, Churchill and his wife decided one night to have a dinner in Monte Carlo. After feasting on soup and pheasant, he chose from the dessert selection a charlotte russe.

After a spoonful, Churchill signaled for a waiter.

When one hurried to the table, Churchill barked, "Pray, take away this pudding—it has no theme."

FIFTH RATE

✢ Sir Cedric Hardwicke, at a reception following one of his plays, was introduced to Churchill. Sir Cedric said, "I was honored to read that you said I was your favorite British actor." Churchill replied, "I was misquoted. What I said was, 'Sir Cedric is my fifth favorite actor. The first four are the Marx Brothers.' "

FIGURE OF SPEECH

⚜ In 1900, the twenty-six-year-old Churchill, after just being elected to Parliament, made a speaking tour of America. In Washington, he was introduced to a majestically endowed woman from Richmond, Virginia, who prided herself upon her devotion to the "lost cause of the Confederacy." Her family were Democrats who had opposed the Republican policy of Reconstruction.

Anxious that Churchill should know her sentiments, she remarked as she gave him her hand, "Mr. Churchill, you see before you a rebel who has not been 'Reconstructed.' "

"Madam," he replied with a deep bow that surveyed her décolletage, "reconstruction in your case would be blasphemous."

FLY-ING HIGH

⚜ In the 1930s, a Liberal Party statesman spoke at a London dinner honoring the League of Nations. The speaker soared to rhetorical heights as he depicted the day when there would be no war amid an era of international brotherhood. Afterward, a listener asked Churchill, who had attended the dinner, what he thought of it.

"Well," he commented. "It was good. It had to be good, for it contained all the platitudes known to man with the possible exception of 'Prepare to meet thy God' and 'Please adjust your trousers before leaving.' "

FOR THE LOVE OF ALLAH!

✢ As prime minister in World War II, Churchill found himself, at a state banquet at 10 Downing Street, seated next to the Iraqi ambassador. After the dinner, Churchill said, "Ambassador, why don't you come back to my study for a nightcap?"

"Mr. Churchill," replied the ambassador, "I can't. I'm a Muslim."

"What, you don't drink? Good God . . . I mean Jesus Christ . . . I mean Allah!"

FRACTURED FRENCH

✢ If "Franglais" has been only recently coined to describe the bastardizing of the French language by English words, Churchill may have been the sire of this hybrid argot. Sometimes his additions to the noble Gallic tongue were even more atrocious than his accent.

During some delicate negotiations at Casablanca, the stubborn Charles de Gaulle denounced an Allied plan to fuse him and his rival, French general Henri Giraud. Churchill, glaring at de Gaulle, delivered this concoction: *Si vous m'obstaclerez, je vous liquiderai!* (If you obstacle me, I will liquidate you!) A bewildered de Gaulle backed off.

FREAK SHOW

✢ Between the wars, Churchill was often exasperated by the impotence of the government in foreign policy.

On one occasion he told the House that as a boy he always looked forward to the London arrival of the American Barnum & Bailey Circus.

"But," added Churchill "there was one show that my nanny would not let me see. She said it was 'too revolting a spectacle for the human eye.' The sideshow was called 'the Boneless Wonder.'

"Now, after thirty-six years, where do I finally find this freak show? Not in the circus, but in the House of Commons, sitting on the front bench—the Boneless Wonder."

GENERAL NUISANCE

✤ When Churchill was president of the Board of Trade in 1909, he was targeted for criticism by Conservative members, whose party he had left. Churchill had this retort ready when one of the senior members of the Opposition attacked his handling of certain ministry matters:

"I admire the martial and commanding air with which the right honourable gentleman treats the facts. He stands no nonsense from them."

GOD WILLING

✤ In the 1930s, Charlie Chaplin visited Chartwell, the Churchill country house. At the dinner table, Churchill asked the movie actor what film project he was next considering.

Chaplin replied, "In all seriousness, I'd like to play Jesus Christ."

Churchill paused and then looked over his spectacles. "Have you cleared the rights?"

GRAVE CONCERN

✤ At Harrow School, Winston did not win any prizes for his behavior and comportment. After a host of infractions, young Churchill was issued an order to report to the headmaster's office.

When the thirteen-year-old Churchill entered the study, Dr. J.E.C. Welldon raised his six-foot frame from his chair and stared down at his subordinate student. With his hands folded behind his back, Welldon intoned, "Young man, I have grave reasons to be displeased with your conduct."

The boy Churchill looked up and replied, with equal solemnity, "And similarly I have grave reasons to be displeased with your conduct."

GRIN AND BEAR IT

✣ As Churchill boarded a splendid yacht at Cannes, he was asked, "Sir Winston, are you looking forward to your Mediterranean cruise?"

Churchill replied, "I always manage somehow to adjust to any new level of luxury without whimper or complaint. It is one of my more winning traits."

HIGH SPIRITS

✣ In 1941, Prime Minister Winston Churchill visited General Montgomery. After a morning session of inspecting the troops, the prime minister offered a nip of whiskey.

Monty refused, pounding his chest with the boast, "I neither drink nor smoke and I'm one hundred percent fit."

Churchill put down his cigar and lifted his glass, replying, "I both drink and smoke and I'm two hundred percent fit."

HUGLESS HERO

✣ It was December 1941 when Churchill came to America for the first time as prime minister. The resolute leader whose defense beat back a Nazi invasion was to Americans the world's most heroic figure at that time.

In Washington he addressed a joint session of Congress. The next day he lit the White House Christmas tree. At such a holiday time, Churchill must have sorely missed his wife, Clementine, and his family.

On Christmas Eve, Diana Hopkins, the nine-year-old daughter of presidential aide Harry Hopkins, heard a knock on her bedroom door. It was the White House butler.

"Miss Hopkins," he said, "the prime minister wants to see you."

Apprehensive about what this great man wanted, she put on her robe over her pajamas and was escorted to the Monroe Bedroom, where Churchill was staying.

The butler knocked and there was a growl of response.

"Miss Hopkins, sir," whispered the butler.

When the door opened, the butler left. Churchill, clad in a green silk robe, held out his arms and embraced the awestruck Diana, saying, "I'm a lonely old father and grandfather on Christmas Eve who wanted a little girl to hug." Then he sent her back to bed.

I DANCED WITH A GIRL WHO DANCED WITH A MAN WHO DANCED WITH A GIRL WHO DANCED WITH . . .

⚜ In Morocco during conferences for the North African campaign, Churchill joined festivities at the British Officer's Club one night.

His eye was caught by an elegant blond lady who sat alone at a table. He asked his secretary, John Colville, why such a beautiful woman was alone. "No doubt her husband," said Colville, "is away on a flying mission. Her husband, I am told, is in the RAF."

"Such beauty," said Churchill, "should never be unattended," and he went over and asked her to dance.

They did, and afterward she quickly vanished. The next day an unsigned note was delivered to his headquarters:

"Prime Minister, thank you. I can tell my grandchildren that I once danced with the greatest man in the world."

The Prime Minister never found out the name of his mysterious dancing partner.

INFORMATION, PLEASE

✣ The eighty-year-old prime minister, in a political debate, was besieged by a Socialist who tested the aged warrior with a series of eight long questions, each beginning with "Is it not a fact . . . ?"

Maintaining his composure, Churchill replied, "The gentleman seems more interested in imparting information than securing it."

"JOHNNY [BULL] REB"

✣ While visiting Williamsburg, Virginia, Churchill was asked what it felt like to be in a city that played such a great role in the Revolutionary War against the English.

"Revolution against the English! Nay, it was a reaffirmation of English rights. Englishmen battling a Hun king and his Hessian hirelings to protect their English birthright. Such a struggle against a German despot is a scene not unfamiliar to English-speaking peoples who twice this century have triumphed over Teutonic tyranny."

LADIES ONLY

✤ In April 1953—just before the coronation of Queen Elizabeth II—Churchill and his wife were invited to Windsor Castle, where in St. George's Chapel Churchill was to receive the Order of the Garter.

Driving from Chartwell to Windsor for the knighting ceremony, Mrs. Churchill felt a call of nature and asked that they stop at a petrol station. As she let herself out of the car to make her way to the restroom marked "Ladies," Churchill called out, "No—Clementine, you can't go now. You're not a 'Lady' yet. You'll have to wait until after the queen awards me knighthood."

"THE LAST SHALL BE FIRST"

✤ Over the course of his lifetime, Churchill was awarded many honorary degrees. At the University of Miami in 1946, after being awarded his doctorate of laws, he made this comment:

"Perhaps no one has ever passed so few examinations and received so many degrees."

LIFTING OF SPIRITS

✤ In 1951, when Churchill returned to 10 Downing Street as Prime Minister, one of the first trips he took was to France to meet with General Eisenhower, the newly appointed head of the Allied NATO command.

At his chateau outside Paris, General Eisenhower entertained his old war comrade at a luncheon. During the luncheon Eisenhower spoke earnestly of the need for more forces. He continued well after the dessert.

Finally Churchill, noticing an ornate credenza behind Eisenhower on which a decanter of brandy stood, said, "Dwight, that's a handsome credenza. Is it Louis Seize?"

Eisenhower—despite a nudge by his deputy general, Alfred Gruenther—said, "I guess it is. It was here when I came," and Ike went on speaking about the need for the enlargement of the British contingent.

Then Churchill interjected, "And that's a splendid decanter on the credenza. Is it Austrian crystal?"

Ike replied, "I suppose, but about this manpower problem—"

Whereupon Churchill again interrupted, saying, "More than manpower, it's morale—and the first thing the supreme Allied commander must do is lift the 'spirits' on that credenza."

LION'S SHARE

❧ In 1942, Churchill was presented a gift of a lion. The lion, named Rota, was placed in the London Zoo. Rota was a favorite of Churchill's and he regularly went to the zoo to share a moment with another creature noted for its roar.

A reporter knowing of Churchill's fondness for his pet did a feature story on Rota, with an accompanying photo of the lion with a gaping mouth in the midst of a bellow.

Back at 10 Downing Street, the complement of staff included a small, slight, bespectacled male clerk who flustered at any Churchill challenge to his work.

One day when the clerk's revised offering was incorrect for the second time, Churchill sent for him.

"You see this picture of Rota. He is hungry for meat. Meat is scarce in wartime Britain. But if you fail for a third time, you will be Rota's din-din!"

The hysterical clerk fled to his cubbyhole and immediately tendered his resignation, saying to his associates that the prime minister had become a raving lunatic.

LITTLE BOY—SHOO!

✢ A cook at Chartwell invited her daughter and grandson to visit her at teatime. In the catching up of conversation, the unwatched boy escaped from the kitchen to wander in the house.

Opening the door of Churchill's study, he marched up to the desk of the prime minister, who was in the midst of drafting a speech.

"Are you really the greatest man in the world?" he asked.

"Of course I am—now buzz off."

LOUSE IN THE HOUSE

✢ In 1947, Hugh Gaitskell, the Socialist Minister of Fuel, in a speech in the House of Commons asked the nation to take fewer baths to save coal. "As a matter of fact, I have never had a great many baths myself."

Churchill, as the Leader of the Conservative Party Opposition, interrupted to say, "When Ministers of the Crown speak like this on behalf of His Majesty's Government, the Prime Minister and his friends have no need to wonder why they are getting increasingly into bad odor.

"I have never asked myself, when meditating on these points, whether you, Mr. Speaker, would admit the word 'lousy' as a Parliamentary expression in referring to this administration, provided, of course, it was not intended in a contemptuous sense but one purely as one of factual narration."

LUCK OF THE DRAW

✢ Twice did Harrovian student Churchill fail his entrance exam into Sandhurst, the British equivalent of West Point. If his third examination paper fell below a score of sixty, he would be forever barred from qualifying for the Royal Military Academy.

In studying for the exam, Churchill decided to concentrate on the geography question that amounted to one fifth of the total grade. He snipped maps of the British dominions out of an atlas and dumped them into a top hat and figured he had time to study only one. Putting his hand over his eyes, he shuffled them around and then picked out one. It was New Zealand.

He traced and retraced the North and South Islands that constitute the empire dominion.

On examination day, the sergeant major who was giving the exam wrote out on the blackboard, "Question #1 (20 points): Draw a map of New Zealand."

As Churchill told it later, "It was like breaking the bank at Monte Carlo. I drew both islands, put in the key cities and river. Then I added the parks, the libraries, and even the tram lines, I got the full twenty points and squeaked into Sandhurst with a sixty!"

"MAD DOGS AND ENGLISHMEN"

✢ In 1940, when the Japanese policy of military aggression threatened the British possessions of Hong Kong, Singapore, and Malaya, the British minister in Tokyo wrote Prime Minister Churchill seeking advice.

Churchill, who personified the English bulldog, wired back. His three-word telegram was a pugnacious snarl—"PESTER—NAG—BITE!!"

MASTER RACE

✣ The story is told that after the British landing defeat in Norway in March 1940, a proposal was made to the First Lord of the Admiralty that sheaths should be provided to all the Royal Marines to protect the ten-and-a-half-inch barrel of their rifles from sweating and then freezing in the Arctic temperatures.

A pharmaceutical company known for manufacturing condoms was asked to provide a test model. In the meantime, Churchill had become prime minister.

In his office, Churchill looked at the box, delivered from the pharmaceutical company, that lay before him.

"Won't do," he muttered. Then he drew one of the cartons out of the box.

He shook his head. "Won't do." Then he opened one of the cartons to extract a packet. "It won't do," was again his pronouncement.

"What do you mean?" remonstrated an aide. "It will sheath the ten-and-a-half-inch barrel."

"There's no labeling," replied Churchill.

"Labeling?" said the confused aide.

"I want a label for every box, every carton, every packet, saying, 'British—Size Medium.' That will show the Nazis, if they ever recover one of them, who's the master race!"

MATE AND SWITCH

✛ In December 1941, Prime Minister Churchill was invited to address a joint session of House and Senate. Churchill, who was half American, told the American lawmakers:

"If my father had been American and my mother English—instead of the other way around—I might have got here on my own."

MISTAKEN IDENTITY

✛ Prime Minister Neville Chamberlain was the apostle of appeasement. Churchill once described him as having "a lust for peace." In 1938 the one-time industrialist from Birmingham, speaking of his peace preparations for his forthcoming trip to Munich to negotiate with Hitler, ended his report on an emotional note: "I cannot remember a time when I was not told stories of Bethlehem, the birthplace of the Prince of Peace."

As Chamberlain paused for breath, parliamentarians could hear Churchill's sardonic mutter "Bethlehem? I thought he was born in Birmingham."

MORALITY TALE

✛ When Churchill first served in the Liberal cabinet, the prime minister was Herbert Asquith and the Opposition leader was Arthur Balfour, head of the Conservative Party.

In an argument with Churchill, a Conservative member, referring to Asquith's extracurricular affairs, said Asquith was "wicked."

"No," said Churchill, thinking of the slippery but chaste bachelor Balfour. "Balfour is wicked and moral. Asquith is good and immoral."

MORNING AFTER

❧ One night in the House of Commons, Churchill, after imbibing a few drinks, stumbled into Bessie Braddock, a corpulent Labourite member from Liverpool. An angry Bessie straightened her clothes and addressed the British statesman.

"Winston," she roared. "You are drunk, and what's more, you are disgustingly drunk."

Churchill, surveying Bessie, replied, "And might I say, Mrs. Braddock, you are ugly, and what's more, disgustingly ugly. But tomorrow," Churchill added, "I shall be sober."

THE MOUSE THAT ROARED

❧ One weekend, while Churchill was resting at Chequers, the official country home of the British prime minister, an aide found Churchill retouching with his brush and oils Rubens's original masterpiece of the ensnared lion being rescued by a mouse.

Like a naughty boy caught in the act, Churchill looked sheepish. Then he muttered in defense, "Poor little mouse! If he were to gnaw the ropes and rescue the lion, he had to be made bigger!"

MOVEMENT IN B-MAJOR

❧ After the war, when Churchill was leader of the Opposition, he was felled by an attack of pneumonia. During the illness, his trained nurse heard him chuckling as she left the room bearing a bedpan. She said, "Mr. Churchill, I don't see anything funny about taking out a bedpan."

Replied Churchill, "It's not you. I'm congratulating myself. It's the first time a movement I've backed has been carried out since the Socialist government came in."

NO CAUSE FOR CELEBRATION

✢ When Churchill visited America in 1941, he was asked to speak at a celebration in Yorktown of that city's role in achieving the victory in the War of Independence.

Churchill replied, "I would be glad to oblige, but not to celebrate!"

NO IFS, ANDS, OR "BUTTS"

✢ After the British deliverance at Dunkirk, Churchill, in the House of Commons, rallied Britain with his most memorable speech.

"We shall fight on the beaches, we shall fight on the landing grounds, we shall fight in the fields and in the streets, we shall fight in the hills. We shall never surrender," he declared.

Then, as the House of Commons thundered in an uproar at his stirring rhetoric, Churchill muttered in a whispered aside to a colleague, "And we'll fight them with the butt ends of broken beer bottles because that's bloody well all we've got!"

NO RUSH JOB

✢ In early 1945, President Roosevelt wrote to Churchill about the agenda for the six-day conference of the big powers at Yalta. FDR saw there was no reason why the plans for establishing the U.N. could not be completed in the conference session.

Churchill was doubtful.

"I don't see any way of realizing our hopes for a World Organization in six days," he wrote FDR. "Even the Almighty took seven."

NOTHING AT STAKE

✛ At the Casablanca Conference in 1942, Churchill and Roosevelt met with Charles de Gaulle, the prickly leader of the French Resistance.

Brendan Bracken, who was in charge of Churchill's arrangements at the hectic three-day meeting, at one point vented his frustrations at his boss.

Churchill replied, "Well, Brendan, you have only one cross to bear. I have a double cross—the double cross of Lorraine."

"The general's problem," sympathized Bracken, "is that he thinks he is the reincarnation of Joan of Arc."

"No, the problem is," concluded Churchill, "my bishops won't allow me to burn him."

NOT "BUY" THE BOOK!

✛ In 1936 when *Gone With the Wind* was the rage, a reporter asked Churchill if he had bought the bestselling novel.

"There is a rule," replied Churchill, "that before getting a new book, one should read an old classic.

"Yet as an author," continued Churchill, "I should not recommend too strict an adherence to this rule."

NOT HUNG UP OVER IT

✛ Once Churchill was sitting on an outside platform waiting to speak to crowds who had packed the streets to hear him. Beside him the chairwoman of the proceedings leaned over and said, "Doesn't it thrill you, Mr. Churchill, to see all those people out there who came just to see you?"

Churchill replied, "It is quite flattering, but whenever I feel this way I always remember that if instead of making a political speech I was being hanged, the crowd would be twice as big."

NOT WORTH THE PAPER ...

✤ A young member of Parliament delivered a speech at a banquet which Churchill attended. Afterward he asked the great orator how he did.

"First, you read the speech. Secondly, you read it badly. Finally— it wasn't a speech worth reading!"

THE OINK AND THE PUSSYCAT

✤ Churchill's endearment for his wife, Clementine, was "pushy-cat." And what did Clementine call him, of all things, but "pig." And sometimes each would call out to the other, respectively, "Meow, meow" or "oink, oink."

Once at 10 Downing Street during the war, both were on the floor on their hands and knees sounding out "Oink, oink" and "Meow, meow" when the Archbishop of Canterbury walked in. He must have thought his prime minister had gone absolutely bonkers!

ONE AT A TIME

✤ During the wartime coalition, Churchill assigned to some La-bourites a few of the more ceremonial but less meaningful ministries. One of these plums was given to the Lord Privy Seal, whose responsibilities included the supervision of state papers. A particular document needed the signature of the prime minister, and the Lord Privy Seal dispatched his young aide to track down Churchill.

Churchill was finally traced to the House of Commons lavatory, where clouds of billowing cigar smoke behind a stall door signaled his presence.

"Mr. Prime Minister," the aide said, rapping on the door, "the Lord Privy Seal requests your signature at once on a document important to the crown."

Churchill, annoyed at being pestered by a man he thought he had carefully shelved, bellowed, "Tell the Lord Privy Seal that I am sealed in my 'privy.' " Then he added, "And I can only deal with one shit at a time."

OPTION RENEWAL

❧ During a speaking tour in Canada, Churchill attended a reception and found himself seated next to a stiff-necked Methodist bishop.

A pert young waitress appeared with a tray of sherry glasses. She offered one to Churchill, which he took, and then one to the Methodist bishop. The bishop was aghast at the alcoholic offer, saying, "Young lady, I'd rather commit adultery than take an intoxicating beverage."

Thereupon, Churchill beckoned the girl. "Come back, lassie; I didn't know we had a choice."

ORATE AND OSCULATE

❧ A reporter once asked the old statesman what in his life did he find were the most difficult tests.

Churchill replied:

"To climb a ladder leaning towards you, to kiss a girl leaning away from you, and third, to give an after-dinner speech."

OVERSHADOWED

✢ Churchill did not get along with Lord Reith, the dour head of the BBC. Reith's gloomy presence was enhanced by his towering six-foot-four frame. Once, when the huge dark-visaged Scot was seated next to the prime minister, Churchill grumbled, "Who will rescue me from this 'Wuthering Height'?"

PAPER THIN

✢ After a cross-country tour of the United States in the 1930s, Churchill was questioned in a Canadian press interview.

"Mr. Churchill," a reporter asked, "Do you have any criticism of America?"

Churchill thought and then replied, "Toilet paper too thin! Newspapers too fat!"

PAPERWEIGHT

✢ As Prime Minister Churchill neared his eightieth year, his innate intolerance for the turgidity of bureaucratic prose grew even thinner. Once at a cabinet meeting, his Minister for Housing asked Churchill if he had studied his memorandum on council rentals. Churchill, lifting the thick booklet in his hands, barked, "This paper by its very length defends itself against the risk of being read."

PATERNAL PRIVILEGE

✢ Churchill once said, "My idea of a good dinner is to dine well, and to discuss a good topic with myself as chief conversationalist." Once after a meal, his son Randolph ventured his own opinion.

Churchill broke in with his own version. When Randolph tried to pick up the thread of his discussion, his father barked, "Don't interrupt me when I'm interrupting."

PEER-LESS

In the 1930s, some ministers in Prime Minister Stanley Baldwin's Conservative government were angling to be made Lords. Ennoblement is not something one can openly work for, since it is awarded by the Crown.

Some of the names whispered about as members who might possibly be named barons included a pair who had been most hostile to Churchill.

A Conservative colleague asked, "Winston, what do you think about peerages for them?"

Churchill replied, "Peerages, no—disappearages, yes!"

"P-G" RATING

The weekend house of the British prime minister is Chequers. The north-of-London Buckinghamshire County residence houses a filming room, which the Churchills often used to enjoy first-run movies.

Churchill would usually ensconce himself in the first row with his miniature poodle, Rufus, sitting on his haunches between Churchill's knees.

One evening in 1952 the Churchills were watching *Oliver Twist*. When Bill Sikes drowns his dog to prevent police from tracking him, Churchill covered his poodle's eyes. "You can't watch, Rufus—I'll explain to you later."

If it was not the British version of "Parental Guidance," it was at least "Poodle Guidance."

"PIECE" AT ANY PRICE

✜ A story is told at Yalta of a reported toast Churchill made to Premier Joseph Stalin. Following President Roosevelt's glowing tribute to the Russian dictator, an aide nudged Churchill to follow suit, but Churchill muttered, "But they do not want peace."

Finally, noting the stare of the Soviet delegation, he rose and said, "To Premier Stalin, whose conduct of foreign policy manifests a desire for peace." Then, away from the translator, he whispered, "A piece of Poland, a piece of Czechoslovakia, a piece of Romania . . ."

PILLAR OF STRENGTH

✜ One Easter Sunday in the 1950s, Churchill and his wife drove out from their Chartwell home to the little Anglican church at nearby Edenbridge.

After the service, the rector shook hands with Churchill and took note of one of his rare appearances. "You're not quite a pillar of the church, Prime Minister."

"No," replied Churchill. "I'm not a pillar, but a buttress. I support it from the outside."

PIMPLY PUT-DOWN

✜ During the 1930s, while speaking out against the Nazi menace, Churchill at a talk to his constituency at Epping was heckled by a youth hardly more than teenaged. At the second of the adolescent's taunts, Churchill replied, "I admire a manly man, and I rejoice in a womanly woman—but I cannot abide a boily boy. Come back in a few years when your cause is as free from spots as your complexion."

PREPOSITIONAL NITPICKING

✤ A priggish civil servant had corrected and returned a Churchill memorandum, pointing out that the prime minister had mistakenly ended a sentence with a preposition.

Back it went to the officious bureaucrat, with this Churchill note appended in the margin:

"This is the sort of pedantic nonsense up with which I will not put."

PRACTICE MAKES PERFECT

✤ In 1926 Churchill as Chancellor of the Exchequer was defending his budget to the House of Commons when Philip Snowden, the former Labourite chancellor, interrupted to accuse his successor of switching positions.

Churchill replied, "There is nothing wrong with change—if it is in the right direction."

Snowden countered, "You are an authority on that."

Churchill rejoined with impish glee, "To improve is to change; to be perfect is to change often."

PRIME TIME

✤ When Churchill returned to 10 Downing Street for the second time, in 1951, there was some criticism about his advanced age. A year later a reporter cornered the seventy-eight-year-old prime minister and asked him if he was going to make his announcement to retire soon.

Churchill growled, "Not until I'm a great deal worse and the empire a great deal better."

PRINCE CONSORT

✦ At a formal banquet in London, the attending dignitaries asked the question, "If you could not be who you are, who would you like to be?" Naturally everyone was curious as to what Churchill, who was seated next to his beloved Clemmie, would say. Would Churchill say Julius Caesar or Napoleon? When it finally came Churchill's turn, the old man, who was the dinner's last respondent to the question, rose and gave his answer.

"If I could not be who I am, I would most like to be . . ." and here he paused to take his wife's hand, "Mrs. Churchill's second husband."

PRIVATE BUSINESS

✦ In the gentlemen's lavatory in the British House of Commons there is a very long urinal with a great row of appliances. One day, Clement Attlee, a leader of the Labour Party, was addressing a spot near the door when Winston Churchill entered and walked all the way to the other end of the room to do his business. Said Mr. Attlee, "Winston, I know we're political opponents, but we don't have to carry our differences into the gentlemen's lavatory."

Churchill replied, "Clement, the trouble with you Socialists is that whenever you see anything in robust and sturdy condition you want the government to regulate it."

PRIVATE ENTERPRISE

✣ Once, out on the hustings, a Socialist was droning on about the accomplishments of socialism. As an example, he cited the increase in general population under the previous three years of the Labourites' administration. Churchill's ears pricked up as he heard the figure and rose to ask the speaker a question.

"Wouldn't the honourable gentleman concede that the last statistic about population is due to private enterprise?"

PURE AND UNADULTERATED

✣ A reporter once asked Churchill what led him into politics.

"Ambition," Churchill replied. "Pure unadulterated ambition."

"What made you stay in politics?" was the next question.

"Anger," Churchill replied. "Pure unadulterated anger."

RAIN CHECK

✣ George Bernard Shaw wired Churchill in 1931, "Am reserving two tickets for you on opening night of my new play. Come bring a friend—if you have one."

Churchill composed the return telegram.

"Impossible for me to attend first performance. Would like to attend second night—if there is one."

RAT RACE

✣ In 1922 Churchill left office when the Liberal Party was virtually wiped off the electoral map in the devasting defeat of that year. Churchill, who had left the Conservative Party to become a Liberal in

1904, now decided to seek election as a Conservative once more. Bids from local Conservative groups did not pour in. When a friend in the Conservative Party told him, "Winston, you can't expect Conservatives to take you back so quickly," Churchill replied, "It is one thing to rat—but another to re-rat."

RAW DEAL

✢ From his Back Bench seat, Churchill during the 1930s made regular attacks against the Conservative government's defense posture. During one of his House of Commons philippics denouncing military unpreparedness, Churchill was challenged on the accuracy of his projected estimates by a Conservative defense minister. In his answer Churchill faulted the scanty information released by the Army Ministry.

Then a heckler from the Conservative ranks yelled out, "Like any bad bridge player the Right Honourable Gentleman blames his cards."

"No," snapped back Churchill, "I blame the crooked deal!"

RECIPROCAL TRADE

✢ During a visit by Churchill to Richmond, Virginia, a memorial sculpture to the wartime prime minister was dedicated. A southern lady of Rubenesque proportions gushed to Churchill when she met him in a receiving line, "Mr. Churchill, I want you to know I got up at dawn and drove a hundred miles for the unveiling of your bust."

"Madam," replied Churchill, gazing at her amply endowed figure, "I want you to know that I would happily reciprocate the honor."

"RED" PALEFACE

✢ Once when Churchill was staying at the White House, Mrs. Roosevelt attacked him for his colonialist views on India.

"The Indians," she charged, "have suffered for years under British oppression."

Churchill replied, "Well, Mrs. R., are we talking about the brown-skinned Indians in India who have multiplied under benevolent British rule, or are we speaking about the red-skinned Indians in America who, I understand, are now almost extinct?"

REINFORCEMENT OF SPIRIT

✢ In his first presentation of the budget in 1925 as the new Chancellor of the Exchequer, Churchill estimated the incoming revenues. When he came to excise taxes on liquor, he paused and reached for a glass—except that it wasn't filled with water but whiskey, and Churchill said:

"It is imperative to fortify the revenue, and this I shall now—with the permission of the Commons—proceed to do."

THE RELUCTANT DRAGON

✢ Late in the 1920s, Churchill was arguing for greater expenditures for defense. His critics, including most of the leading British statesmen, argued that peace with aggressive dictators could be ensured by the League of Nations and negotiation in good faith.

To the St. George Society, Churchill described how St. George today would attempt to save a maiden from the dragon.

"St. George would be accompanied, not by a horse, but a delegation. He would be armed not with a lance, but by a secretariat."

Churchill continued by adding that St. George "would propose a conference with the dragon—a Round Table conference—no doubt that would be more convenient for the dragon's tail.

"Then after making a trade agreement with the dragon," Churchill continued, "St. George would lend the dragon a lot of money."

Finally, Churchill summarized, "The maiden's release would be referred to the League of Nations of Geneva and finally St. George would be photographed with the dragon."

RENDEZVOUS WITH DESTINY

✢ On Churchill's eightieth birthday, a reporter encountered him as he left his Hyde Park residence in London.

"Mr. Churchill, do you have any fear of death?"

"I am ready to meet my Maker," Churchill replied, and then he added with a twinkle, "But whether my Maker is prepared for the great ordeal of meeting me is another matter."

RINGMASTER

✢ World War I began when Churchill was the First Lord of the Admiralty. As Churchill predicted, Germany invaded France through neutral Belgium. While Belgium contemplated surrender, the British military command recommended a diversionary operation in Belgium to keep German troops that were deployed in Belgium away from France.

Churchill, donning the uniform of his honorific position as Elder of the Trinity House, commandeered five Piccadilly buses and arrived in Belgium with two thousand untested Royal Marines.

In his braided eighteenth-century uniform and tricornered hat he looked like a cross between a circus ringmaster and Napoleon. But the grandiose spectacle of the costumed Churchill deploying red London buses to and fro across Belgium directing the setting up of defensive positions so lifted the morale of the Belgians and captured the attention of the Kaiser's Imperial Army that the Germans dispatched from the French front.

SCAREDY CAT

✢ Churchill had a pet black cat whom he named Nelson after the famous British admiral, the victor at Trafalgar.

During a nighttime air-raid bombing, Nelson cowered under the bed in Churchill's underground annex next to 10 Downing Street.

"Nelson," Churchill roared, quoting the words of Shakespeare in *Henry V*, "summon the spirit of the tiger." Nelson heeded not. Then Churchill added, "Think of your namesake—no one named Nelson slinks under a bed in a time of crisis."

Nelson then reappeared and sniffed, but returned, at the sound of the next siren, to safety under the bed.

SCRATCH MY BACK

✢ In 1898, Churchill first met during a country weekend a little-known writer named W. Somerset Maugham. After Sunday dinner, Churchill asked Maugham to take a walk with him. They excused themselves from the guests.

During their stroll, Churchill said, "Maugham, I've been observing you and listening to you. You are intelligent, you express yourself well. You will go far. Now I would like to have an understanding with you. If you will not talk against me, I will not talk against you."

SCREAMING EAGLE

✣ On March 2, 1946, former prime minister Winston Churchill met President Truman at Union Station in Washington. They were to board the private presidential train, the *Ferdinand Magellan*, which would take the presidential party and its famous British guest to Fulton, Missouri, where Churchill would deliver his "Iron Curtain" address.

Before entraining, Harry Truman pointed out the presidential seal that hung on the railing of the caboose car. The car would be the platform on which the president and his guest might deliver remarks at various stops along the way.

"Mr. Churchill, look at the seal. I had it changed. The eagle used to face the arrows—but I had it switched to face olive branches. What do you think?"

Churchill, remembering the warning he was about to deliver against the military threat of the Soviet Union, replied, "Mr. President, with all due respect, I would like to have the American eagle's neck on a swivel—to turn to the olive branch or to the arrows as the situation demands."

SHORT AND SWEET

✣ Churchill and his friend F. E. Smith formed before World War I the "Other Club." Unlike most London clubs, it did not draw its clientele from any common avocation, profession, or political party. Instead, it recruited a motley membership—and invited raucous banter and debate.

One of its charades was to invite a member to give an extemporaneous talk on a subject told to him just after he was called to the head of the table.

The acting chairman called out "Mr. Churchill," and on the placard flashed to him and the eight other members was the word SEX.

Churchill looked at the placard, paused, and intoned deliberately, "Sex . . . it gives me great pleasure."

Then Churchill sat down.

SIMPLE TASTES

❦ The director of the Plaza Hotel in New York was in a stew. Churchill was scheduled to stay at his palatial hotel off Central Park, but nothing was known of Churchill's preferences in food and drink.

In desperation he called the British Embassy in Washington, saying to the clerk who answered, "Mr. Churchill will be staying at our hotel. I am asking if I may find out his preferences and—"

Before he could finish, he was switched off and he heard a growl that sounded something like a "Yes."

"I am the director of the Plaza Hotel inquiring about Mr. Churchill's tastes—"

In a guttural lisp, the voice on the phone answered, "Mr. Churchill is a man of simple tastes—easily satisfied with the best!"

SOCK IT TO HIM

❦ Just after his shattering defeat in 1945, word came to Churchill from Buckingham Palace that King George wanted to invest him with the highest of knightly honors—the Order of the Garter.

Churchill replied, "Why should I accept the Order of the Garter from my sovereign when I have already received from the people . . . the 'Order of the Boot'?"

SO WHAT'S NEW?

✤ In January 1960, a reporter for the *London Standard* approached Churchill at a reception.

"Sir Winston, what is your comment on the prediction made the other day that in the year 2000, women will rule the world?"

"They still will, will they?" was Churchill's grunted response.

SOLOMONIC WISDOM

✤ In 1942, not the least of the frictions between American and British officers in North Africa were the rules in the Officers' Mess. The Americans, who liked to down their whiskey highballs before sitting down to dine, proscribed alcohol during the meal. The British regulations, however, forbade drinking before dinner, but then they would enjoy at the table their claret and burgundy.

Churchill announced his solution to the impasse. "Before dinner we British will have to defer to the American rules. But at the table, you Yanks must abide by the British regulations."

As Churchill raised his own glass of whiskey and soda at the Mess bar, he observed, "I hope this felicitous arrangement for the fraternity of Anglo-American relationships will be accepted in good 'spirits' by all."

STRIPPED CLEAN

✤ On the day of the British general election in 1922, Churchill was convalescing from an appendicitis operation. The attack had stopped short his campaigning. In addition to this handicap, the Liberal Party, to which he belonged, had been badly split by dissenting factions. When the returns came in, he was in a hospital listening on a new contraption called the "wireless." He shook his head sadly at the reports of the landslide defeat and murmured, "All of a sudden, I find myself without a party, without a ministry, and without a seat, and even . . . without an appendix."

STUD SERVICE

✣ Among the varied pursuits of this Renaissance man, Churchill bred racehorses. His hottest prospect was a black stallion named Colonist II, which was eventually entered in the Derby in 1949.

Just before the race, Churchill, as was his custom, entered the stables and proceeded to stroke the horse's black mane. Then the voice that had inspired England during its time of supreme challenge tried to rouse his steed to feats of glory.

"Now, Colonist—if you win today, it will be your last race. Think of it—the rest of your life in stud. Just imagine those verdant pastures replete with lissome nubile fillies ready to attend to every need."

In the race that afternoon, alas, Colonist's efforts placed him far out of the running.

When Churchill was asked what happened, he replied, "Poor Colonist, I told him of the agreeable female company he could expect if he won—but then he couldn't keep his mind on the race!"

STUDIO "ADDRESS"

✣ During the war, Churchill one day left the House of Commons to hail a cab to go to Shepherd's Bush, the site of the BBC Studio. Churchill was scheduled to deliver an 8:00 P.M. address to the nation.

At his wave, a cab came to a halt and Churchill gave the address of the BBC studio.

"Sorry, sir," was the reply. "I have the radio on and I want to hear the prime minister's address."

Churchill, delighted with his response, slapped a five-pound note in his hand. "Driver, I have to get there fast."

The driver replied, "Frig the bloody prime minister, guv'nor, what's that address again?"

SUPER SERVICE

✢ Late in his life, Sir Winston took a cruise on an Italian ship. A journalist from a New York newspaper approached the former prime minister to ask him why he chose to travel on an Italian line when the *Queen Elizabeth* under the British flag was available.

Churchill gave the question his consideration and then gravely replied.

"There are three things I like about Italian ships. First, their cuisine, which is unsurpassed. Second, their service, which is quite superb." And then Sir Winston added, "And then—in time of emergency—there is none of this nonsense about women and children first."

A SWITCH IN TIME

✢ When Churchill in 1904 first met Clementine Hozier, the granddaughter of the Earl of Airlie, it was almost love at first sight.

Churchill at that time was a member of the Cabinet in the Liberal government. Miss Hozier, although a bright young woman knowledgeable in politics, was employed as a governess. Churchill had been introduced to the comely Clementine briefly on a previous occasion. Before dinner was served, he found time to check the seating at the dinner table and changed the place cards so that he would be seated next to the object of his dreams.

TAKING THE IN-LAW INTO YOUR OWN HANDS

❖ Sarah Churchill, Churchill's oldest daughter, married Vic Oliver, a music-hall comedian. At a family dinner at Chartwell, Oliver, who brought along a guest, tried to draw out his famous father-in-law from one of his periodic sullen moods. Oliver didn't strengthen the tenuous relationship with Churchill by his habit of referring to him as "Popsy."

"Popsy," he said, "who, in your opinion, was the greatest statesman you have ever known?"

"Benito Mussolini," was the unexpected reply.

"What? Why is that, Popsy?" said a surprised Oliver.

"Mussolini is the only statesman," grumbled Churchill, "who had the requisite courage to have his own son-in-law executed."[1]

TALL ORDER

❖ Churchill, a guest of President Truman in 1952, one evening was entertained at a dinner in the high-ceilinged salon in the *Williamsburg*, the presidential yacht, as it cruised down the Potomac River. After dinner, Churchill, while enjoying his postprandial cigar and brandy, posed this question to his scientific aide, Lord Cherwell:

---

[1] Count Ciano, the former Italian foreign secretary who married Mussolini's daughter, was shot in 1942.

"Prof, if all the wine and spirits that I have drunk in a lifetime was poured into this salon, do you think it would reach the ceiling?"

Cherwell, who had been a mathematics professor before being commandeered by Churchill to serve as his atomic energy consultant, drew out his ever-present slide rule and made his calculations.

"Prime Minister, if all the alcohol you have consumed in your life were to occupy this room, I estimate that it would only reach the level of your eye."

A disappointed Churchill muttered, "As I gaze at the ceiling and contemplate my seventy-five years, my only thought is—how much left to do and how little time to do it."

TEETOTALING SHIPS

✣ As First Lord of the Admiralty in World War I, Churchill reportedly was asked by a temperance group leader to reconsider the Royal Navy's practice of christening ships by breaking a bottle of champagne across the bow.

"But madam," replied Churchill, "the hallowed custom of the Royal Navy is indeed a splendid example of temperance. The ship takes its first sip of wine and then proceeds on water ever after."

THEIR TURN

✣ Not since Caesar had Italian military prowess been the subject of admiration. In 1940 Prime Minister Churchill was handed a report that Italy had entered the war on the side of Germany. He read it and chortled.

"Why are you laughing, Prime Minister?" asked his aide.

"It's only fair, don't you see?" replied Churchill. "They were on our side the last time."

THERE GOES THE NEIGHBORHOOD

⚜ During World War II Churchill was occasionally the guest of Franklin Roosevelt at his home in Hyde Park, New York. Once while the President was driving the Prime Minister around his Hudson River estate, Squire Roosevelt observed proudly, "You know, Winston, my Dutch ancestors were among the very first settlers here—in what was then called Nieuw Amsterdam."

Churchill, whose American grandmother claimed to be one-eighth Iroquois, retorted, "But, Franklin, it was my ancestors who greeted them."

THRILL OF THE CHASE

⚜ Churchill's habitual tardiness was a source of continual vexation to his friends. On one occasion, a cabinet colleague and his aides settled in their train compartment to await Churchill, who was scheduled to make an inspection tour of a Midlands defense industry.

Just as the train began to chug away from Liverpool Station, a panting prime minister clambered abroad.

"Winston," chided the Cabinet secretary, "why do you always leave so late?"

"Well, it's only fair," replied Churchill, "I like to give the train a sporting chance."

TO ERR IS HUMAN

⚜ During the wartime coalition, the president of the Board of Trade was the austere Calvinist Sir Stafford Cripps. Cripps both tithed and teetotaled. He was even a vegetarian. His only concession to pleasure was smoking cigars.

This habit too he swore off during the war, when he announced at a rally that he was now giving up cigars as a salutary example of sacrifice.

Prime Minister Churchill, who was seated on the same platform, leaned over to a colleague and whispered, "Too bad—it was his last contact with humanity."

TOO MUCH SNOOZE

✥ Churchill during the war was a tireless worker. He regularly toiled writing his memoranda (*Action This Day*) in the wee morning hours. In sharp contrast, General Bernard Montgomery, Britain's top army commander, retired early and kept regular hours.

Once when the two were conferring in Montgomery's headquarters, the general glanced at his watch at 10:00 P.M. and observed, with a suggestion of weariness, "Prime Minister, it's past my bedtime. Why don't we call a halt?"

Reluctantly, Churchill agreed. In the morning when they met again, the prime minister inquired, "Do you feel rested now?"

"No," answered Montgomery. "I have a headache. I didn't sleep enough."

"I have a headache, too," declared Churchill. "I slept too much!"

TO "PEE" OR NOT TO "PEE"

✥ Churchill was a lover of Shakespeare. In fact, he knew by heart all the tragedies. In 1953, Richard Burton was playing Hamlet at the Old Vic. The prime minister attended a performance, taking his seat in the first row.

Whenever Burton as Hamlet would speak, he could hear the echo of Churchill's growl mouthing the same lines.

Burton tried to step up his pace. Churchill did likewise. Then he slowed his delivery. Churchill again followed suit. Then Burton tried omitting whole lines, only to hear Churchill's roar of protest from the first row.

At the end of the first act, Burton retired to his dressing room. His dresser told him, "I think the old man has left." Just as Burton replied, "Well, that's a relief," Churchill walked in and said, "Prince Hamlet, may I use your lavatory?"

A TOWER-ING FIGURE

✤ In 1943, a stumbling block to Allied plans for the North African campaign was the proud and lofty Charles de Gaulle. The head of the Free French in London was balking at negotiations with the Vichy regime in Algeria. When General de Gaulle insisted upon flying to Algeria, Churchill denied his request.

A furious de Gaulle raged, "So I am a prisoner—the next step will be shipping me off to exile to the Isle of Man [a British wartime center for internment]."

"No, *mon general*," replied Churchill, "for someone as eminent and distinguished as you, it can only be the Tower of London."

TURN OF THE WORM

✤ Violet Asquith, the irrepressible daughter of Prime Minister Herbert Asquith, found a kindred spirit in Churchill, who served in her father's Cabinet.

Once, in a flight of philosophic gloom, she turned to her dinner partner and said, "Winston, in terms of infinity, we are cosmic dust—we are just worms."

"Perhaps, Violet," Churchill replied, "but I am a glowworm."

HIGH-JACK-ING HARRY

✤ A year after he was voted out of office in 1945, Churchill visited the United States. In Washington, he boarded the private train *Ferdinand Magellan* with President Harry Truman.

During the night on the train, Churchill was introduced to the game of poker by Truman and his military aide, General Harry Vaughan.

Losing steadily, Churchill looked at the big pot with only a pair of jacks in his hand.

Staring back and forth at Truman and Vaughan, he grumbled aloud, "Should I wager Britain's precious sterling on a couple of knaves?"

The double bluff worked and Truman and Vaughan folded.

TURNING THE TABLES

✤ As a boy, Churchill found the study of Latin an ordeal. Once the schoolmaster asked the ten-year-old Churchill to decline the words for "table." Churchill proceeded to give *mensa* in all its cases: nominative, genitive, dative, accusative, and ablative.

The master reprimanded him. "You have forgotten the vocative '*O Mensa*' when you would address a table."

"But," Churchill replied, "I don't ever intend to talk to tables."

TWISTING THE LION'S TAIL

❧ On the train to the speech at Fulton where he would deliver his "Iron Curtain" address, Churchill complained to President Truman of the absence of wine at dinner in the lounge car. "Why do you Americans stop your drinking when you sit down to dinner?"

Truman replied, "If you English didn't drink at the dinner table, perhaps you wouldn't have to ask for these loans."

Churchill answered, "Why do you keep twisting the loan's [lion's] tail?"

UNEXPERT OPINION

❧ One day in the House of Commons, a Socialist poured out abusive words against Prime Minister Churchill. Churchill remained impassive, almost bored.

When the harangue was over, Churchill rose and said, "If I valued the opinion of the honourable gentleman, I might get angry."

UNHAPPY COINCIDENCE

❧ Richard Crossman was a left-wing political gadfly who often crossed swords with Churchill, both in his regular newspaper column and on the floor of the House.

Churchill once punctured an attack by Crossman by remarking, "The honourable gentleman is never fortuitous in the coincidence of his facts with truth."

UNMAIDENLY CONDUCT

❧ Novelist and humorist A. P. Herbert was elected to the House of Commons in 1938. His first speech in Parliament glittered with his raucous wit. Someone asked Churchill, "What did you think of Herbert's maiden address?"

"It was no maiden address," retorted Churchill. "It was a brazen hussy of a speech! Never did such a painted lady of a speech parade itself before such a modest Parliament."

UN-PRIME BEEF

✤ When Churchill suffered his devastating defeat in 1945, he was urged by many to retire and give up his leadership of the Conservative party.

In retirement he could visit cities in the United Kingdom and the Commonwealth and bask in the honors extended him by a grateful people. Churchill was adamant in his refusal:

"I refuse to be paraded on exhibition like an old bull at county fairs admired for his past prowess."

WHITE MEAT

✤ At a Richmond, Virginia, reception in Churchill's honor, cold fried chicken was served along with champagne. His hostess was a woman in whom endowment for motherhood was doubly manifest. Churchill approached the buffet table. "May I have a breast?"

Whereupon his hostess, a lady of Victorian sensibility and Virginia gentility, gently chided him. "We southern ladies use the term 'white meat.' "

The next day, Churchill ever a gentleman, sent his hostess flowers. Actually, it was a corsage. Attached to it was a card—"Winston Churchill, M.P." On the other side he had scrawled a note:

"I would be most obliged if you would pin this on your 'white meat.' "

THE WORLD IS NOT MY OYSTER

✢ Near the close of his life, Churchill, though still a member of Parliament, rarely left his 28 Hyde Park residence. Thus, Olive French, the banquet head at the Savoy Hotel, was surprised by a call from Churchill's secretary requesting a table for Sir Winston and Lady Churchill at the Grill at 8:00 P.M. The maître d' and staff were assembled to ensure that the service and food would be impeccable.

The great man in his eighty-sixth year opened his repast with oysters and champagne. His soup course was a pea purée complemented by a glass of fine sherry. Then he attacked a poached Turbot washed down with some Pouilly Fuissé.

For the meat entrée, he dined on a rare slab of roast beef with carrots and broiled potatoes. With the beef he drank a bottle of burgundy.

Dessert was crême brulée with a glass of Madeira. And finally Stilton and port. After coffee he concluded the night with a Romeo & Juliet Havana and a snifter of Napoleon brandy.

Around noon the next day Olive French heard that a call awaited her from Churchill's secretary. She picked up the phone expecting to hear kudos for her orchestrated culinary performance.

"This is Anthony Montague-Brown, Churchill's parliamentary secretary. Sir Winston believes that there was something wrong with the oysters served last night at the Savoy. He is feeling somewhat indisposed today."

YO HO HO AND A BOTTLE OF RUM

✢ Churchill in 1912 became the youngest First Lord of the Admiralty in British history. He quickly set out to reform the Royal Navy. He eliminated the dreadnoughts and replaced them with more mobile battleships and converted the fleet from coal to oil. Then he sent into early retirement many mossback admirals. One such admiral expostulated to Churchill, "You are scuttling the traditions of the Royal Navy!"

"I'll tell you," retorted Churchill "what the traditions of the Royal Navy are—rum, sodomy and the cat-o'-nine-tail lash."

Z AS IN SNOOZE

✣ An American general once asked Prime Minister Churchill to look over the draft of an address he had prepared.

The speech was returned with a note saying, "Too many passives and too many zeds."

Later the general asked Churchill to clarify his criticism.

Churchill replied, "Too many Latinate polysyllablics like 'systematize,' 'prioritize' and 'finalize.'

"And then the passives. What if I had said—instead of 'We shall fight on the beaches'—'Hostilities will be engaged with our adversary on the coastal perimeter'?"

Milestones

1874, NOVEMBER 30	Birth of Winston Churchill at Blenheim Palace.
1884, APRIL 17	Enters Harrow.
1893, JUNE 28	Enters Royal Military College, Sandhurst, as cavalry cadet.
1894, DECEMBER	Graduates from Sandhurst.
1895, JANUARY 24	Death of Lord Randolph Churchill.
1895, APRIL 1	Commissioned as lieutenant in the Fourth Queen's Own Hussars.
1895, JULY 3	Death of Mrs. Everest, Churchill's nanny.
1895, NOVEMBER 30	Observes fighting during visit to Cuba.
1896, OCTOBER 3	Arrives in India and settles down in military cantonment at Bangalore. Reads avidly.
1897, SEPTEMBER 4	Takes part in fighting on the northwest frontier of India.
1898, SEPTEMBER 2	Takes part in charge of Twenty-first Lancers at Omdurman in the Sudan, the last cavalry charge in British history.

1899, JULY	Presents himself as Conservative candidate at by-election in Oldham and is defeated.
1899, OCTOBER 14	Sails to South Africa as war correspondent for the *Morning Post*.
1899, DECEMBER 13	Escapes from prison in Pretoria, South Africa.
1900, OCTOBER 1	Elected Conservative member of Parliament from Oldham.
1901, FEBRUARY 14	Takes his seat in the House of Commons.
1901, FEBRUARY 18	Makes his maiden speech in Parliament.
1901, MAY 13	Attacks the army appropriations by the Conservative government Cabinet minister.
1904, MAY 31	Joins Liberal party.
1905, DECEMBER 9	Becomes Undersecretary of State for the Colonies.
1908, APRIL 24	Joins Asquith's cabinet as president of the Board of Trade.
1908, SEPTEMBER 12	Marries Clementine Hozier.
1910, NOVEMBER 8	Welsh mine strike quelled.
1911, JANUARY 3	Battle of Sidney Street. Churchill as Home Secretary had to quell the riot.
1914, JUNE 28	Assassination of Archduke Franz Ferdinand at Sarajevo.
1914, AUGUST 1	Orders mobilization of the Royal Navy.
1914, AUGUST 4	Great Britain declares war on Germany.
1914, OCTOBER 3–6	Churchill in Antwerp, organizes Belgian defense called the Antwerp Circus.

1915, JANUARY 3	With Lord Kitchener, proposes naval and military attack on the Dardanelles.
1915, MAY 28	Resigns as First Lord of the Admiralty following Dardanelles defeat.
1915, NOVEMBER 19	Commands Second Battalion of the Grenadier Guards in France; later commands a battalion of the Sixth Royal Scots Fusiliers.
1917, JULY 16	Becomes Minister of Munitions in Lloyd George's government.
1918, NOVEMBER 11	Armistice signed.
1919, JANUARY 15	Becomes Secretary of State for War and Minister of Air.
1921, FEBRUARY 15	Becomes Colonial Secretary and begins negotiating settlements in Ireland and Middle East.
1921, JUNE 29	Death of his mother, Lady Randolph Churchill.
1921, AUGUST 23	Death of his daughter Marigold.
1922, OCTOBER	Defeated; out of Parliament first time since 1910.
1922, NOVEMBER	Buys Chartwell Manor near Westerham, Kent.
1924, SEPTEMBER	Switches to Conservative Party and is elected as Member from Epping.
1924, NOVEMBER 7	Becomes Chancellor of the Exchequer in Stanley Baldwin's government.
1933, AUGUST	Warns against German rearmament.
1938, OCTOBER	Attacks Munich Agreement.

1939, SEPTEMBER 1	Hitler invades Poland.
1939, SEPTEMBER 3	Britain and France declare war on Germany. Churchill joins Chamberlain's government as First Lord of the Admiralty.
1940, APRIL 9	Germany invades Denmark and Norway.
1940, MAY 10	Churchill becomes prime minister. Germany invades Holland and Belgium.
1940, MAY 13	In his first speech as prime minister, Churchill offers the House of Commons nothing but "blood, toil, tears and sweat."
1940, AUGUST 10– SEPTEMBER 15	Battle of Britain. The RAF repels the German Luftwaffe.
1941, AUGUST 10	Atlantic meeting with President Roosevelt on board *Prince of Wales*. Atlantic Charter signed two days later.
1941, DECEMBER 7	Japan attacks Pearl Harbor and Singapore.
1943, NOVEMBER 28– DECEMBER 1	Meets Stalin for the first time at the Teheran Conference.
1944, JUNE 6	D-Day. Allied invasion of Normandy.
1945, FEBRUARY 4–12	Yalta Conference.
1945, APRIL 12	Death of President Roosevelt.
1945, MAY 8	VE Day. Unconditional surrender of all German armed forces.
1945, JULY 26	Defeated in general election, he resigns as prime minister.
1945, AUGUST 14	VJ Day. Japan surrenders.

1946, MARCH 5	Delivers "Iron Curtain" speech at Westminster College, Fulton, Missouri.
1951, OCTOBER 26	Becomes prime minister for second time.
1953, DECEMBER 10	Awarded Nobel Prize in Literature.
1955, APRIL 5	Resigns as prime minister.
1963, APRIL 9	President Kennedy declares him an honorary citizen of the United States.
1964, JULY 28	Presented with vote of thanks by House of Commons.
1965, JANUARY 24	Dies in his home at Hyde Park Gate in his ninety-first year on the seventieth anniversary of the death of his father, Lord Randolph.

Books and Writings
by Sir Winston Churchill

The Story of the Malakand Field Force, London: Longman, 1898.
The River War (2 vols.), Longman, 1900.
Ian Hamilton's March, Longman, 1900.
Savrola, Longman, 1900.
London to Ladysmith, Longman, 1900.
Liberalism and Socialism, Pamphlet, 1908.
The People's Rights, Hodder & Stoughton, 1909.
The World Crisis (6 vols.), Thornton Butterworth, 1923–1931.
My Early Life, Thornton Butterworth, 1930.
Thoughts and Adventures, Thornton Butterworth, 1932.
Marlborough: His Life and Times (4 vols.), Harrap, 1933–1939.
Great Contemporaries (biographical essays), Thornton Butterworth, 1937.
Step by Step (collected newspaper articles), 1936–1939.
Arms and the Covenant (1928–38 speeches), Harrap, 1938.
The War Speeches (1940–1945), (3 vols.), London: Cassell, 1946.
Painting as a Pastime, Oldhams, 1948.
The Second World War (6 vols.), Boston: Houghton Mifflin, 1948–1954.
Postwar Speeches (5 vols.), Cassell, 1948–61.
A History of the English-Speaking Peoples (4 vols.), London: Cassell, 1956–1958.

Bibliography

Churchill, Randolph S. *Winston S. Churchill. Vol. 1: Youth, 1974– 1900,* Companion (in two parts). Boston: Houghton Mifflin, 1966.

Churchill, Randolph S. *Winston S. Churchill. Vol. 2: Young Statesman, 1901–1914,* Companion (in three parts). Boston: Houghton Mifflin, 1967.

Gilbert, Martin. *Winston. Churchill. Vol. 3: The Challenge of War, 1914–16,* Companion (in two parts). Boston: Houghton Mifflin.

Gilbert, Martin. *Winston S. Churchill. Vol. 4: The Stricken World, 1917–22,* Companion (in three parts). Boston: Houghton Mifflin.

Gilbert, Martin. *Winston S. Churchill. Vol. 5: The Wilderness Years, 1923–37,* Companion (in three parts). Boston: Houghton Mifflin.

Gilbert, Martin. *Winston S. Churchill. Vol. 6: Finest Hour, 1937–41.* Boston: Houghton Mifflin.

Gilbert, Martin. *Winston S. Churchill. Vol. 7: War Years, 1942–45.* Boston: Houghton Mifflin.

Gilbert, Martin. *Winston S. Churchill. Vol. 8: Never Despair, 1946–65.* Boston: Houghton Mifflin.

Soames, Mary. *Clementine Churchill.* Boston: Houghton Mifflin, 1979.

McGowan, Norman. *My Years with Churchill*. London: Souvenir Press, 1958.

Moran, Lord. *Churchill*. Boston: Houghton Mifflin, 1966.

Sykes, Adam, and Sproat, Iain. *The Wit of Sir Winston*. London: Leslie Frewin, 1965.

Nel, Elizabeth. *Mr. Churchill's Secretary*. New York: Coward McCann, 1958.

Halle, Kay, ed. *Irrepressible Churchill*. Cleveland: World Publishing, 1966.

Howells, Roy. *Churchill's Last Years*. New York: David McKay, 1965.

Coote, Colin, R., ed. *A Churchill Reader*. Boston: Houghton Mifflin, 1954.

Manchester, William. *The Last Lion: Winston Spencer Churchill— Visions of Glory*. Boston: Little Brown, 1983.

Manchester, William. *The Last Lion: Winston Spencer Churchill— Alone*. Boston: Little Brown, 1988.

Eade, Charles, ed. *Churchill by His Contemporaries*. New York: Simon & Schuster, 1954.

James, Robert Rhodes, ed. *Speeches of Winston Churchill, 1897–1963* (8 vols.). New York: Chelsea House, 1977.

Churchill, Sarah. *A Thread in the Tapestry*. New York: Dodd, Mead, 1967.

Taylor, Robert Lewis. *Winston Churchill: An Informal Study of Greatness*. Garden City, NY: Doubleday, 1952.

Bonham-Carter, Violet. *Winston Churchill: An Intimate Portrait*. New York: Harcourt Brace, 1965.

Adler, Bill. *The Churchill Wit*. New York: Coward McCann, 1965.

Moir, Phyllis. *I Was Winston Churchill's Secretary*. New York: Wilfred Funk, 1941.

Cowles, Virginia. *Winston Churchill: The Era and the Man*. New York: Harper & Brothers, 1953.

Thompson, Walter H. *Assignment Churchill*. New York: Farrar Straus & Young, 1965.

Sherwood, Robert E. *Roosevelt and Hopkins*. New York: Harper & Brothers, 1948.

Mendelssohn, Peter de. *The Age of Churchill: Heritage and Adventure 1874–1911*. New York: Knopf, 1961.

McGurrin, James. *Bourke Cockran*. Boston: Scribner's, 1948.

James, Robert Rhodes. *Churchill: A Study in Failure*. New York: World, 1970.

Cook, Donald. *Charles De Gaulle*. New York: Putnam, 1983.

Pilpel, Robert H. *Churchill in America (1895–1961)*. New York: Harcourt Brace Jovanovich, 1976.

Weidhorn, Manfred (intro.). *India—Defending the Jewel in the Crown*. Contoocook, N.H.: Dragonwyck, 1990.

Humes, James C. *Churchill: Speaker of the Century*. Stein and Day, 1980.

Bibescu, Princess. *Winston Churchill: Master of Courage*. London: Robert Hale Ltd., 1957.

Langworth, Richard, ed. *Finest Hour*. Hopkinton, N.H., 1981–present. Quarterly journal of the International Churchill Societies.

Czarnomski, F. B., ed. *Wisdom of Winston Churchill*. London: Allen & Unwin, 1956.

The International Churchill Societies

Founded in 1968 and active on three continents, the International Churchill Societies are five independent, nonprofit educational organizations which work together, in the words of their charter, "to keep the memory green and the record accurate," so that future generations will never forget Sir Winston Churchill's contributions to the political philosophy, culture, and literature of the English-speaking peoples. ICS/USA is now creating a Churchill Center for the Study of Statecraft in Washington, D.C. Aside from its educational and research functions, the Churchill Center aims to create a computerized index to everything Churchill wrote and spoke, a quantum leap forward for Churchill scholars.

Friends of ICS are from all walks of life—academics, statesmen, students, professionals, nonprofessionals, collectors, bibliophiles, teachers—interested in some aspect of Churchill and his career, not merely in Churchill as the symbol of victory of war, but as a symbol of culture, humor, principle, optimism, pride in country, and faith in Western civilization.

The society's quarterly journal, *Finest Hour,* often touches on Churchill's political philosophy and its relevance to problems of the present. Society members unite not to worship Churchill but to study his myriad experience, as a sure guide to our own cultural and national lives—and, in particular, they study the wit and wisdom which enabled him to survive the storms that have rocked our troubled century. ICS has done much to preserve our memory of Churchill, promoting publication of a dozen long-forgotten Churchill books and itself publishing numerous books and monographs, including forgotten works by Churchill himself, which are usually sent free to current members.

By becoming a member of the Churchill Societies, you not only assist in the pursuit of these worthy goals, but open an avenue of personal contact with like-minded people through many local chapters and national or international events.

ICS United States
1847 Stonewood Drive
Baton Rouge, LA 17847

ICS Canada
130 Collingsbrook Boulevard
Agincourt, Ontario, Canada M1W 1M7

ICS United Kingdom
29 High Street
Shoreham
Seven Oaks, Kent, England TN4 7TD

ICS Australia
8 Regnans Avenue
Endeavour Hills, Victoria, Australia 3802

ICS New Zealand
5 Basilton Close
Bucklands Beach
Auckland, New Zealand

Index